Lecture Notes in Computer Science 741

Edited by G. Goos and J. Hartmanis

Advisory Board: W. Brauer   D. Gries   J. Stoer

Bart Preneel René Govaerts
Joos Vandewalle (Eds.)

Computer Security and Industrial Cryptography

State of the Art and Evolution

ESAT Course
Leuven, Belgium, May 21-23, 1991

Springer-Verlag

Berlin Heidelberg New York
London Paris Tokyo
Hong Kong Barcelona
Budapest

Series Editors

Gerhard Goos
Universität Karlsruhe
Postfach 69 80
Vincenz-Priessnitz-Straße 1
D-76131 Karlsruhe, Germany

Juris Hartmanis
Cornell University
Department of Computer Science
4130 Upson Hall
Ithaca, NY 14853, USA

Volume Editors

Bart Preneel
René Govaerts
Joos Vandewalle
Departement Elektrotechniek, Katholieke Universiteit Leuven
Kardinaal Mercierlaan 94, B-3001 Heverlee, Belgium

CR Subject Classification (1991): C.2.0, D.4.6, E.3-4, G.2.1, K.6.5

ISBN 3-540-57341-0Springer-Verlag Berlin Heidelberg New York
ISBN 0-387-57341-0 Springer-Verlag New York Berlin Heidelberg

© Springer-Verlag Berlin Heidelberg 1993
Printed in Germany

Typesetting: Camera-ready by author
Printing and binding: Druckhaus Beltz, Hemsbach/Bergstr.
45/3140-543210 - Printed on acid-free paper

Preface

The ESAT Laboratorium of the Department of Electrical Engineering at the Katholieke Universiteit Leuven regularly organizes a course on the state of the art and evolution of computer security and industrial cryptography. The first course took place in 1983, the second in 1989, the third in 1991, and the fourth course is scheduled for 1993.

The ESAT course is intended for both researchers and practitioners from industry and government. It covers the basic principles as well as the most recent developments. Because of our background and because of the relevance, the emphasis lies on cryptography without forgetting the most important topics in computer security. We try to strike the right balance between basic theory and real life applications, between mathematical background and juridical aspects, and between recent technical developments and standardization issues.

During our 1991 course Walter Fumy suggested editing the text of the speakers into more formal written documents. All speakers were invited to submit their contributions, and almost all of them responded positively with an excellent text. We feel that the result – complementary to text books and conference proceedings – can be very interesting for those interested in cryptography and computer security. We would like to thank the authors for their careful preparation of their contributions.

Leuven, Belgium   B.P, R.G, and J.V.
1993

Contents

Section 1

Introduction

Trends in the Fight Against Computer-Related Delinquency

Prof. Dr. B. De Schutter

Director Center for International Criminal Law
Vrije Universiteit Brussel

1 Characteristics of the Phenomenon

The grasp of information technology upon almost all societal activities is an indisputable and irreversible fact. Transfer of data, information, knowledge or know-how has undergone with the technological wave a profound change in its form, speed and distance coverage. This mutative effect can certainly be beneficial to society in all its components (economic, strategic, intellectual, cultural).

It seems, however, that the margin between use and abuse is rather narrow. Even if criminality related to information has always existed, the intervention of the computer with its above-mentioned characteristics of time, volume and place, leads to the risk of a criminal activity, the nature of which might be different from the more classical information crimes. To look into the phenomenon, its size frequency and profile, will lead to the necessary conclusion for the need of policies, which may be necessary to effectively combat this anti-social behaviour.

In that exercise one encounters a number of difficulties. A first one concerns already the definition of computerdelinquency. According to the purpose for which it is needed, one can work with a more criminology-oriented definition, describing the deviant pattern from the sociological angle, or could need a more precise terminology when introducing the illegal act as crime in the penal arena, then requiring precise material and moral elements in the definition. Avoiding the multitude – and the nuances – of definitions of the expert writers [1], there is much merit in the OECD working definition, referring to "any illegal, unethical or unauthorized behaviour relating to the automatic processing and/or the transmission of data" [2], since the answer to computer criminality is likely not to be limited to an exercise of criminal law drafting alone. However, the danger, of such an extensive "opening definition" is that it allows a somewhat overqualification of incidents, in which the computer does not play any instrumental role at all. Some demystifying and relativation has to be done to bring the phenomenon back into real proportions, avoiding the sensationalism of the media.

Theft of microprocessors is not a computercrime, even if their capacity increase as a result of new technologies drastically modified their economic value. But even then, the magnitude of the information technology criminality should not be underscored.

Scarcely a day passes without any newspaper-clip on computer-crime or fraud. Television picks up the item and visualizes the "hacking" techniques. The difficulty, however, is to bring those individual findings into some global figures and clear indicators. This seems, especially in our countries, to be too delicate, if not impossible.

There is a sphere of reluctance and unwillingness in communication of incidents. Banks, insurance companies or any other potential victim are not easily communicative on losses occurred through computer interventions. Image-loss, indirect consequences such as the thrust of the customers or the competitive position, all push towards secrecy and discretion. The simple anonymous transfer of information for statistical purpose to official instances, even international ones, is objected to; the interference of judicial authorities is considered as "counter-productive" ?.

Most known cases come in the daylight through indiscretion, erroneous behaviour of the criminal himself or when insurance companies oblige the client to do so before refunding any loss. Besides, some countries are more communicative than others [3]. For sure one can state that figures are incomplete, that guesses be considered with care and that we only know the top of the ice-berg, whether that means 1% as to the FBI, or 15% as to the very experienced Stanford Research Institute.

Since a few years a considerable number of official bodies or professional circles are, showing interest in gathering valuable information. All of it should be read with a critical eye, since under- or overscoring is likely. Nevertheless, figures are impressive and worthwhile to be recalled: SRI mentions 100 million $/year in the U.S., the FBI makes two billion dollars out of it [4]. For Europe, an interesting estimate is the one of the *Association Internationale pour l'Etude de l'Assurance*, which comes up with six billion dollars loss for Europe in 1988. A U.K. survey by the *Local Government Audit Inspectorate* led in 1984 to 80% of 320 interviewed firms having been victim of a computerfraud [5], while for 1985 four major British banks budgeted £85 million against computer frauds [6]. The French CLUSIF reports a yearly amount of nearly 8 billion FF of Voluntaristic or accidental damages.

It is not the purpose of this paper to recall the spectacular and classical examples such as the Equity-funding [7] or Security Pacific Bank [8], or Memorial Sloan Kettering Cancer Institute [9], the French ISOVER Case [10] or the CLODO activities [11], or many other [12] but it may be important to recall that not all incidents are linked to economic interest as such, but may equally concern health, privacy, morality or state strategic survival.

If the total size of computer abuses is substantially high, though not full-proof, it has also been proven that these totals concern a limited number of victims. Concentrating the losses upon few leads to the conclusion that the average gain of such crime is a hundred times that of the average classical hold-up, while the average time for the "discovery of the discovered" seems to be counted in years, not in months [13].

To be added to this picture is the great potential of the transborder dimension of information technology, whereby the physical presence of the actors upon the foreign territory is no longer necessary. This internationalization of this criminality adds a new dimension to the task of society in reacting against this phenomenon.

As to the actors themselves, they seem roughly to fall into two major groups: on one hand the computer-freaks, the youngsters trying to hack systems for fun, competition, challenge; whizkids or wargamers, i.e. "short-pants criminality", not necessarily with a clear criminal intent; on the other hand, wilful criminality by hackers or employees within the system, often highly qualified and technically skilled, often acting from within, abusing the hi-tech and jargon oriented "state in the state" position of the EDP-sector.

In conclusion on the characteristics of the phenomenon one can say that computers, whether used for simple data storage or retrieval, word processing, business activities, banking, electronic fund transfer, electronic mail, health care, R & D, defence systems, ..., are vulnerable to attack by experts or freaks, young or old, acting from within or without the area of operation of the machine, with results to be estimated with a mutative scale difference, since time, space or volume have no longer a limitative effect.

As to the different possibilities for misuse, – even if they can probably be technically described in a uniformed way – writers have identified several areas of incidents, but fail to bring them back in a uniform classification [14]. This harmonization need is now attempted through the channel of international bodies [15].

Roughly seen a categorization can be brought down along the following lines:

manipulation of data: input of false data, alteration, erasure, deterioration and/or suppression of stored data or programs with fraudulent intent.

data espionage: unlawful collection or acquisition and/or use of data.

computer sabotage: leading to the destruction or disruption of soft- or hardware. Extensively this may include any hindering of the functioning not only of a computer but also of the telecommunication system. Today this includes the phenomenon of viruses and worms.

unauthorized access or interception of a computer and/or telecommunications system with infringement of security measures.

program piracy with the infringement of the exclusive right of the owner of a protected computer program with the intent to exploit commercially the program. The same can be said of a protected chip or microprocessor.

Even if differences of labeling may occur under various initiatives [16], the major phenomena are clearly covered by the above list. As will shown below, not all countries accept the criminalization of all of these acts and the conditions of applicability are even more diversified.

2 The Computer Threat

It would be erroneous to overscore as well as to underestimate the vulnerability of the computer society. It is clear that from the angle of victimology, three major targets groups can be detected.

2.1. the **individual** becomes the weakest link in the new information technology era, not only from a sociological and economic point of view (loss of job security through robotics, word processing, etc...), but equally from the angle of legal protection (privacy).

2.2. the **economic targets** are also rather interesting: banks, insurance companies, corporations of all nature become more and more vulnerable, especially where networks are more and more flourishing, telecommunications more and more used, but still hardly protected, and the cost effectiveness of certain protections not yet shown. Direct and indirect losses will be substantial and the defence of the law too much of an after-the-harm reparation.

2.3. the **sovereign state** itself, who faces a so-called erosion of sovereignty when noticing that many raw data can and will leave the country for economic decision-making abroad (e.g. with the multinationals), the state having no insight in the departure of raw data or return of information and loosing impact on economic or financial decisions taken outside its operating or influence zone and nevertheless having to cope with the possessors of data and/or information.

The key finally becomes not so much the technology itself. It is mainly instrumental to a far more important target which needs protection: the legal interest violated, whether the individual human life, the survival of an economic entity or the independence of the nation itself. The technology adds something, be it speed, massification of data or transfrontier communication. It emphasizes or amplifies, without necessary creating new forms of criminality. Thanks to it, information radically grew in importance and with it all the values attached to it, whether intangible or not. Much value has to go to the notion of *information related criminality* or even asset protection management, in which, besides information, image protection and physical securization become equally important. The massive presence of computers and other devices in the whole flow of information in our society at all levels (international, state, firm or individual) may ultimately lead to an infiltration into the totality of the legal field, the criminal, as well as the civil, administrative, economic or constitutional one. The call for full re-assessment of the whole of the law to make the legal system respond better to the problems of new information technology is real.

Looking into the legislation of mainly the industrialized nations, one notices that in various countries answers have been formulated or are in the process of being formulated [17]. Their responses differ both because of the underlying legal system and of their appraisal of what computer-crime means to them today as a threat. Even if in the more regulated field of privacy protection the reference frame exists with the OECD guidelines [18] and the Council of Europe Convention for the Protection of Individuals with regard to Automatic Processing of Personal Data [19], national laws may show diversified implementation norms or techniques [20]. One has the feeling that national – if not nationalistic – approaches prevail, taking the territorial – thus national – character of the criminal law as a starting point. So is the response to the threats of economic recession or sovereignty erosion.

While information criminality has an important transborder facet and data will be easily send and handled abroad, the need for a more global, uniform or harmonized approach is not always perceived or accepted. A first and maybe not optimal trend, therefore, is the *all too national instead of cooperative* response to the phenomenon.

The efforts of the Council of Europe in the criminal field, or of the EEC in such areas as micro-processor protection [21] or data vulnerability as such [22], should receive priority attention and national legislatures should adjust to them quickly.

3 Trendsetting in Fighting Computer-Related Criminality

Out of the above findings, one must conclude that a valid response to the phenomenon requires a holistic approach, of which the three layers would be:

1. the security issue is a threefold one: it requires technical answers, managerial/organizational ones and legal responses.
2. within the legal sphere, different branches of law have to intervene (criminal law, civil law, intellectual law, labor law, etc...)
3. within the subsets of the law the international cooperation or coordination is indispensable.

3.1. The issue of *information security* cannot be addressed by the law only. Even if criminal sanctions or damage allowance have besides their reparation effect, an educational and preventive effect, it nevertheless is also true that the intervention of legal mechanisms mostly occurs at moments when the harm is already done and the incident consumed. Prevention prevails over repression. To that effect, the tackling of this issue requires an integrated approach of technicians, economists, behaviorists, organizational managers and lawyers. The responsibility of computer firms is involved to the extent that they ought to voluntarily accept minimum security standards or at least make their clients aware of the vulnerable aspects of their computerization and require them to take sufficient starting security measures in relation to issues such as physical integrity of premises, access controls and authentication procedures, the possibility or necessity for back-ups and contingency planning.

The software industry has to continue the search into security software and the protection of expert systems. Management and economists should develop more efforts towards risk analysis and cost/benefit appraisal within a given environment, foresee an appropriate organizational scheme adapted to the information flow structure; behaviorists and industrial psychologists have to look into the personnel hiring and training issues. Lawyers will have to play their part in prevention (e.g. software protection, contract provision, employee contracts, insurance policies) and in the elaboration of specific legislative initiatives in the repressive reply.

3.2. Today most countries do not have a cohesive information law. This obliges us to a search into different sectors so as to reach partial answers. Before focussing upon the criminal law, it may be interesting to look into them in a brief manner.

The first area in which a generally accepted minimum solution was reached is the one of the *protection of personal* data. The centralization of data and the networking of databanks change radically the image a classical file could give of a given person. Besides the OECD guidelines and the Council of Europe Convention, a number of countries now possess specific legislation in this respect, including criminal provisions [23]. Within the European Commission a draft directive circulates in the light of the 1992 internal market (COM(80) 314 def. SYN.287). Other countries trail behind like Belgium. In that case only general provisions of constitutional or criminal nature or the European Human Rights Convention can help.

A second area is the one on *software protection*, via intellectual property law. The question whether one should start form the copyright notions, the patent law concept or seek for a *sui generis* solution has not yet found a definite answer, even though within the international organizations such as WIPO the former solution seems to have gained substantial support [24]. Specific penal provisions can of course be built into these laws. Another area is the one of *labor law*, where most legislations foresee norms on the protection of industrial secrets, trade secrets or secrets related to the relationship employer/employee. Equally operational are legislations on *unfair competition acts* by employers during or after their contractual relationship [25]. In the *law of contracts*, secrecy or non-disclosure clauses can be build in, with provisions for fines in case of non-respect. *Contractual responsibility* rules can play in cases of unlawful use of information [26].

To be complete, one has to mention the *communication legislation*, very often not adapted to the new transfer of information technique, and legally often linked with a monopoly-situation and non-responsibility. In countries where deregulation is introduced (USA, Japan, U.K.), competition may lead to the use of security improvement as an argument in commerce. A breakthrough in the non-responsibility area seems the more urgent move to make, such as France did to introduce PTT-liability for gross negligence [27].

Finally, the legal spectrum also offers the *insurance law* as a protective barrier, once more to cope with casualties, i.e. after the incident. Nevertheless, a good management of a computer-incidents insurance portfolio can be part of a valid answer to criminal situation. Above all, it will need a careful cost/benefit analysis, and will probably have to be linked with a security level audit.

3.3. The extent to which transborder data flows open the risk to transborder criminality leads inevitably to the conclusion that valid answers depend on a harmonized, if not unified, approach by the industrial world. Since free flow of information is the basic rule, this flow must be protected and the international networks secured against criminal attacks. Comparable substantive law should be elaborated, while the international instruments in the field of cooperation in criminal matters be reviewed to be adapted to this specific criminality, a.o. to allow transborder evidence admissibility, based on computer-outputs. All to different legislations or non-acting countries will lead to "information heavens" or "telecom crime paradises", endangering not only the international informational market, but also the economic or private or strategic interests of a given country or person, and thus potentially leading to national protectionist measures, certainly counterproductive in the light of the ultimate goal. The efforts at the level of the Council of Europe to achieve a minimum level of consensus on which acts should be considered criminal, deserve utmost support, at a time where most countries are in the process of preparing legal reforms, either specifically or of the penal code as a whole. When criminals prevail themselves of the advantage of the new information technology to internationalize their acts, there is no reason to remain above all in a nationalistic or territorial limitation in trying to formulate the answer.

4 The Criminal Law Scene

4.1 The Policy Issue

When analyzing the rather scattered legislations on computer-related criminality, it is noticeable that there are important methodological approaches between national laws. Computer-criminality is new to some of them, others, like he U.S., "benefit" from a few decades of experiences and incidents. This leads to diversified general policies, at least for the time being.

One tendency is to consider the area of computer-criminality not as specific or totally new. Crimes against information or against property or privacy have always existed. Even if the new informative technologies throw some particular features in the arena such as volume, speed or distance, it would by no means justify a new legal concept. The computer is to them only an instrument in the commission of a traditional crime, such as theft, embezzlement, violation of trade secrets, etc... No need for legal action in the criminal field would be necessary, emphasis would be upon the civil actions for damages, while the criminal judge would do with existing definitions, eventually going as far as some extensive interpretations. This attitude seems to be limited today to a few countries, which seemingly have not been affected by the phenomenon or, at least, in which no major court activity in computer-crime is noticeable [28].

It is our contention that few, if no industrialized country will be left over in this category, as all nations will be facing serious challenges to the existing laws and the pressure for concerted action a.o. in the European context is strengthening.

The other reaction is to realize that new measures are inevitable. Therein, one can distinguish those who prefer a low profile adaptation, i.e. the analysis of existing concepts, testing their applicability to computer-related situations and, if needed, to take this dimension into account. This can then be done through amending the actual provision [29]. Others wish to enact clearcut new incriminations either as a specific bill [30], or as new provision or even as a new chapter in the penal code [31]. It has to be noticed, at the same time, that many countries are in the process of reviewing the whole of their penal code, which is certainly an excellent opportunity to include at an appropriate place the necessary provisions relating to information technology crimes [32].

In conclusion it seems correct to state that a large majority of concerned countries, together with international organizations such as the OECD or the Council of Europe, are well aware of the necessity to act at legislative level, even though with variable intensity. As will be shown in the following analysis, many states have indeed already taken initiatives or are in the process of doing so.

4.2 The Analytical Survey

The analytical survey of existing laws, drafts, loopholes and problems is not an easy task. Like many other scholars, we have the benefit of the outstanding expertise of Dr. Ulrich Sieber, who together with Martine Briat, was responsible for the survey conducted under the auspices of the OECD's ICCP [33]. The present analysis rests inevitably upon the same material and cannot be considered as exhaustive as the leading publications referred to. As in the OECD we start from the classical five-fold categorization: manipulations, espionage and piracy, sabotage, unauthorized use and unauthorized access. For once, the reversed order will be used, each time reaching a higher degree of criminality. To take the unlawful access and use as a starting point may be justified through the fact of their not so obvious association with the "crime" notion, their rather high frequency and potential danger, while at first glance, they belong to the least protected expressions of the phenomenon.

Unauthorized Access of Computer- and Telecommunication Systems. Notions such as "computerhackers", "whizkids", computer-time theft are already familiar. As of today there is no general penalization of this activity. Some countries have a specific legislation [34]. Analogies may be drawn from articles incriminating the entrance into one's property with false or unlawfully obtained keys or wiretapping of conversations over PTT installations. In the field of privacy protection an occasional provision may be found punishing unauthorized access [35]. In some countries wiretapping of computer communications is punishable (Canada 178-11 Criminal Code) (Belgium Telecom. Law 1930) (U.K. Interception of Communication Act 1985). The Swedish privacy act (1973) includes a provision applicable if no other incrimination can be applied. So does the German Second Law for the prevention of economic crime (1986).

Others, like the French provisions or U.S. proposals [36] go all the way towards the inclusion of such a provision. It must be stressed, however, that such provision should be – and mostly is – conditioned with several elements such as:

- a specific intent (knowingly, without color of right)
- the violation of security measures
- a special request by the victim.

Often criminal prosecution will be waived if the perpetrator informs the victim of the act and indicates the loopholes in the system.

Unauthorized Use of Computer- and Telecommunication Systems.
Most countries do not provide a specific provision on unlawful use (furtum usus). Sometimes, one can rely upon unlawful use of somebody's property, which would more point to the hardware use. This would be possible under Danish, Finnish or English law. In other countries, concepts such as theft of electricity might be applicable (Belgium), while others require the abuse of specific objects, such as cars or bicycles (Netherlands, Germany). Considering this diversity and the rising number of incidents of this nature, the experts both at OECD and the Council of Europe opted for the introduction of specific provision in the minimum model system. Initiatives were already taken at individual levels. Canadian (Criminal Law Amendments Act 1985) and American (Counterfeit Access Device and Computer Fraud and Abuse Act 1984) initiatives have already come through, while the guidelines for national legislatures of the Council of Europe puts the unauthorized use in the so-called optional list.

In the light of this consensus trend, uniform requirements would be a preferable goal. Again one may include:

- specific intent
- infringement of security measures
- intent to cause harm or another form of computer-crime (loss, e.g.).

Such a provision on "furtum usus", if made specifically for information technology issues, requires a clear definition to distinguish between information-processors which should remain outside the scope (wrist watches, pocket calculators) and the real targets, while emphasis should go upon the functions performed and not upon the technological assets, since the latter will be subject to continuous evolution [37].

Finally, it has to be mentioned that such unauthorized use will often occur within the frame of an employment relationship or of a service contract. This indicates that much can be achieved through clear formulation of rights and duties in the contractual or organizational area, and also through security awareness initiatives in DP environment.

Computer-Sabotage. If one considers in this the destruction and/or damaging of computercenters, data or other items linked with the computer, it is clear that this concept goes beyond the physical "terrorism" against corporal elements, but also concerns intangibles such as the data or programs themselves. Phenomena such as viruses and worms resort under this concept. This latter part is mostly not covered by notions such as property damage, vandalism, malicious mischief, since information can e.g. be erased without damaging or destroying the physical carrier. Therefore, countries, in which specific computer-crime law exists or is in preparation, do foresee either a specific comprehensive provision on this issue (American state laws e.g.), or an adaption to the traditional concepts (e.g. the Canadian new sections in the criminal provision on "mischief": mischief in relation to data). Austria, France, Denmark, West Germany, etc..., seem to go for specific computer sabotage provisions, as does the Council of Europe. It clearly indicates that besides the classical protection of tangible property, in one way or another the introduction of penal protection against damage to data or programs is to be suggested. Again, we would plead for a rather high threshold, including:

- specific intent
- detailed description of acts (destruction; damaging, rendering useless, meaningless or ineffective)
- eventually, aggravating levels can be introduced if the target is an essential element in public administration or an economic enterprise.

Computer-Espionage. The major targets to be protected here are the computer-stored data, the special protection to be offered to computer programs and, recently the special protection of computer chips. If it is clear that the illegal appropriation of one's property is perceived as a crime and is covered by many existing provision such as theft, embezzlement, larceny, the specificity here relates to the fact that some of the targets are not of a physical nature, but constitute "intangibles", not covered by these provisions. A basic discussion related to this concerns the legal status of information.

If no proprietary rights are possible, can it then be subjects to "espionage" ? The protection of data stored in a computer system can eventually be looked upon from the *traditional property law angle*. The major problem of the intangible nature of information is sometimes explicitly solved by including express reference in the law (U.K. Theft Act 1968) (Australia) (USA) (Luxembourg draft). Others rely upon notions such as extending the idea of theft of electric impulses, even though electricity is a tangible (hold a wire and you feel it), or assimilating because of the economic values involved (Dutch and Belgian case law). Fundamentally, we can follow the Canadian Sub-Committee on Computer Crime, when opting against the property approach. The reasons are to be found in the above-mentioned aspects, namely tangible property or intellectual public good; traditional property/intellectual property; theft of tangible/intangible. A specific provision is preferable. Other linkages can be found in the *trade secret and unfair competition* law, where many countries foresee criminal provisions within their trade secrets law (West Germany, Switzerland, Austria, Greece). Others only cover partial aspects (e.g. fabrication secrets in Belgium, France, Luxembourg) or rely mainly on civil damages remedies. For the US a recommended Uniform trade secret Act has been adopted by a series of states. The U.K., Canada and Australia have not many penal provisions available, but are in the process of elaborating appropriate responses. So are the Scandinavian countries. This trend deserves support. The balance to be found, however, is here also between the legitimate right of the "owner" or "developer" to have his economic values in data or programs protected and the right of society to have ideas and discoveries accessed by anyone. The transborder dimension of information transfer should add even more to the difficulty of phrasing appropriate provisions, while the specificity of some informations (military, privacy, hi-tech know-how) or of some "detainers" (government officials, police officers, doctors, ...) equally can lead to separate or special rules. Should there be a "informational secrecy" as extension of the classical "professional secret"?

The way this provision should be foreseen can thus raise basic theoretical issues as to the status of the data which are intercepted. Anyway the interception or appropriation of data form part of a broader range of abuses, namely the attack against the integrity of computer- or telecommunication systems. This concerns more the right to undisturbed exchange of data than the consequences itself of acts of espionage.

It would, therefore, be interesting not to cover the data or programs as such, but to search for the penal protection of the integrity of computer access of it. As conditions could be foreseen: the intent to harm.

As to the additional protection of computer programs, leaving aside the unsolved problem of the intellectual property priority of copyright over patent law or/and a sui generis solution [38], the main trend towards the copyright provisions should be followed in a spirit of harmonization, together with a strengthening of the penal sanctions in them, as was done in Italy (law of 1981), Sweden (1982), Finland (1984), West Germany (Copyright Amendment Act 1985 or the U.K. (Copyright Amendment Act 1982).

Computer-Manipulations. This is considered as the modification of data to change the processing and/or the output in order to obtain a change in information or at the expected consequence. In the latter case, one is back into the "property" issue, (fraud, e.g.) with all its difficulties; in the former, forgery is the major available notion. As to fraud, the deception of a computer to meet the requirement that a "person" be deceived, seems problematic. Breach of trust is either limited to qualified persons or also requires a physical transfer of specific objects. Forgery is based upon visually readable documents, humanly understandable. Solutions de lege lata seem indispensable and are already available or under way. New laws are to be found in Sweden (Swedish Data Act 1974), the U.S. (Credit Card Fraud Act 1984) (Counterfeit Access Device and Computer Fraud and Abuse Act 1981), Canada (Criminal Law Amendment Act 1985), Denmark (Data Kriminaliteit Law 1985), West Germany (Second Law for the Prevention of Economic Crime 1986). The Council of Europe expert report lists computer-related fraud and computer forgery among the "hard-core" offences to be covered by all member states [39]. Work is done in the Netherlands, Luxembourg and Belgium. Consensus thus seems reached as to the necessity to act in this sector. Requirements should be a special intent (to obtain an advantage, or to harm) and a formulation in terms of functions and not in terms of todays technology.

4.3   The Transborder Issues

One of the more likely aspects of the phenomenon is its transborder potential. The elaboration of networks, the development of international telecommunications and the presence of a "multi-nationals" oriented economy certainly affect the traditional patterns of information transfer.

This carries consequences to be located in the international criminal law sphere, more particularly those of the penal jurisdictional competencies and the cooperative mechanisms between sovereigns. Answers have to be found to questions such as the localization of the crimes, the territoriality or extra-territoriality of them, the character of the crime (immediate, continuous, collective, . . .), the applicability of the cooperation structures (such as extradition, minor assistance, transfer of proceedings), the police-cooperation, the evidence issue when computer elements are included. Pluri-national incidents are likely to occur with the presence of things such as SWIFT networks, electronic mail, international airline reservation systems, etc. . .

As to the *competence-issue*, the theory of ubiquity may receive a new perspective, whereby the situs of the act, its instrumentality situs, the situs of the potential consequence and the one of the actual effective consequence are and difficult to locate and more diversified than the traditional "shot over the boarder" example.

Considering the non bis in idem principle, a clearer delimitation or at least classification of competencies could become indispensable. It again points to the necessity of harmonized legislations. This "international connectivity" throws new light upon concepts dating from the "before the computer" area.
In the cooperation issue, the problem of double criminality requires once more

a common approach. Elements of distant complicity or co-authorship require response. What also about the effect of certain additional measures imposed as a sanction, such as the interdictions to use data or programs collected or obtained in violation of criminal law provisions. How does the notion of rogatory commissions apply to evidence stored in a foreign database, having an intangible character or/and being accessible from front-ends in the requesting state. How is seizure and restitution of data conceivable between two states. Many are the questions raised, few are yet the answers. The work in the Council of Europe did not lead yet to some specific ones [40].

4.4 The Procedural Issues

As for the transborder situation, a number of problems may occur in the domestic sphere. The most important issue seems here to be the admissibility of computer records as evidence. Most continental law countries have given much power to the criminal judge in the free evaluation of introduced evidence. It could be that no problems arise, even though the problem of authenticity of the evidence may play. In the common law countries, computer evidence may be regarded as "hearsay evidence", basically inadmissible.

Exceptions are made or in the make, such as the U.K. Police and Criminal Evidence Act (Bill S-33). Requirements of accuracy, knowledge of the existence of the automated system and its proper use, complementary or to be supplemented by other proof may be retained.

5 Conclusion

The world of new information technology is one of the most evolutive ones. The somewhat mutative effect of certain of these inventions equally affects the legal components of societal adaptation to them. But law is not knowledgeable for quick responses and immediate flexibility. Especially criminal law should be preserved from an all too hasty reply to timely phenomena. There is a need for a minimum of stabilization of acts or attitudes felt as a danger to society, a sort of confirmation of the discovery of new anti-social behaviour and the clear creation of a sufficient consensus for penalization of it. The computer abuse area has now reached the confirmation phase: facts are clear, continuous and increasing in number and inventiveness. The telecommunications area is now part of the criminal scene, maybe not fully in the open because of the technical unawareness of the victims or their attitude of overdiscretion, but equally vulnerable. The time to respond is there, if we do not wish the phenomenon to grow unharmed, considering the loopholes in the law and the legal vacuum in the transborder aspects of it. Concerted action seems to be the only efficient one, either through conventional way or, at least, through the search for common thresholds. The work of the OECD and the Council of Europe should be regarded as the guiding trends, allowing coherent law-making activity in national parliaments. The balance between overcriminalization and the actual status of underlegislation still has to be found in many countries. To build upon a broader perspective than the national frontiers and to benefit from international expertise in the field seem to be cornerstones for effectiveness. The challenge is real, the social duty to respond to it is also within the hands of the legal profession.

References

1. It seems that every major writer in the field handles its own terms. See:
   SCHJØLBERG, *"Computers and penal legislation"*, Oslo, 1983, p. 3;
   SOLARZ, *"Computer technology and computer crime"*, Stockholm, 1981, p. 23.
   The computer is sometimes the instrument or target of an act (VON ZUR MUHLEN, *"Computer Kriminalität"*, Berlin 1972, p. 17); The specific purpose (PARKER, D.B., *"Computer Abuse Assessment"*, Washington, 1975, p. 3); Bequai only goes for "part of larger forms of criminal activity: white collar crime" (BEQUAI, *Computer crime*, Lexington, 1978, p.1).
2. OECD-ICCP, *Computer related crime - analysis of legal policy*, Paris (1986), p. 7.
3. See U.K.: A. NORMAN, *"Computer Insecurity"*, London, 1983 and K. WONG & FARQUHAR, W., *"Computer related Fraud Casebook"*, BIS Applied Systems, Ltd., Manchester, March 1983, 106 p.
   See also: Australia's information at the Caulfield Institute of Technology, Computer-Abuse Research Bureaus (CIT-CARB); Japan, National Police Department, *"White paper on computer crime"*.
4. Within SRI, D. Parker's publications are the more important ones:
   D.B. PARKER, *"Crime by computer"*, New York, Charles Scribner's and Sons, 1976, 308 p.
   D.B. PARKER and S.B. NYCUM, *"Computer abuse"*, U.S. Department of Commerce, Springfield, NTIS, 1973, 131 p.
   D.B. PARKER, *"Computer Security Management"*, Prentice Hall, 1981, 308 p.
   D.B. PARKER, *"Fighting Computer Crime"*, New York, Charles Scribner's and Sons, 1983, 352 p.
   Drs. J.C. VON DIJK R.A., *"Computercriminaliteit"*, Ars Aequi Libri (ser. Strafrecht en criminologie, dl. 3), p. 203, 1984; Ph. JOST, "Les pillards d'ordinateur défient le FBI", VSD, 27 jan. 1983, p. 10.
5. Local government Audit Inspectorate – Department of the Environment, *Computer Fraud Survey Report*, Crown copyright, July 1981, 33 p.
6. Scottish Law Commission, Consultative Memorandum no. 68, *Computer Crime*, (March 1986), p. 1.
7. L.J. SEIDLER, F. ANDREWS and M.J. EPSTEIN: *"The Equity Funding Papers, the anatomy of a fraud"*, New York (J. Wiley and Sons, Inc., 1977).
8. "10.2 Million $ theft may yield profit for victim", EDCAPS, Jan. '79, p. 11-12, Aug. '79, p. 14-15.
9. Whiz-kids managed to establish a link with a private network of the General Telephone Electronic Telenet.
   S. HERTEN, "Computercriminaliteit: er is nog toekomst in de toekomst", *Humo*, 26 Jan. 1984, nr. 2264, p. 32.

10. S. O'DY, "Informatique: La chasse aux perceurs de codes", *L'express*, 9 March 1984, p. 74-75.

11. *L'express*, 26 April 1980, *L'informatique nouvelle*, Sept. 1980, p. 25.

12. *Science et Vie micro*, Dec. 1983, p. 52.
   J. BLOOMBECKER, "International Computer Crime: Where Terrorism and Transborder Data Flow Meet", *Computers & Security*, 1982, p. 41-53.

13. G. LEVEILLE, "Data base security systems, protecting the source", *Today's Office*, Oct. 1982, p. 62.
   D.B. PARKER, *"Computer abuse Assessment"*, Washington D.C. (National Science Foundation), 1975, p. 32.
   B. ALLAN, "The biggest computer frauds: lessons for CPA's", *The Journal of Accountancy*, May 1977, p. 52-62.

14. U. SIEBER, *"Gefahr und Abwehr der Computercriminalität"*, Betriebs-Berater 1982, p. 1433; A. NORMAN, *"Computer Insecurity"*, London (1983); US Department of Justice, *Computer Crime, Criminal Justice Resource Manual*, Washington, 1979.

15. We refer to the work of the OECD-ICCP ad hoc group of experts on computer-related criminality (1984-85) and the Council of Europe's Select Committee of experts on Computer-related Crime (CDPC-PC-R-CC) (1985-1989).

16. See e.g. the 8 categories of the Scottish Law Commission, including a.o. eavesdropping on a computer. Supra note 6, p. 18.

17. As to the criminal law initiative, one can only refer to the excellent study of Miss. M. Briat and Prof. U. Sieber leading to the OECD publication on computer related crime. OECD-ICCP, *Computer related crime – analysis of legal policy*, Paris 1986, p. 79.

18. OECD Guidelines governing the Protection of Privacy and Transborder Flows of Personal Data, 1980.

19. 1981, in force Oct. 1, 1985, ratified by France, Austria, Denmark, Germany, Luxembourg, Norway, Spain, Sweden, U.K.

20. U. SIEBER, *The "international handbook on computer crime"*, West Chicester (1986) p. 94 et seq.

21. Council Directive of Dec. 16, 1986 on the legal protection of topografics of semi-conductor products, COM (86).

22. See e.g. the call for proposals in the field of data processing (85/C 204/02) O.J. 13.8.85, no. C 204/2, including the item of a data protective guide for European users.

23. E.g. Denmark: (*the Danish Private registers etc. Act No. 293, 8 June 1978 and the Danish Public Authorities' Registers Act, No. 294 of 8 June 1978*).
   France: (*La loi relative á l'informatique, aux fichiers et aux libertés, no. 78-17, 6 January 1977, B.G.B.I., I S 201*).
   Luxembourg: (*Loi du 31 mars 1979 réglementant l'utilisation des données nominatives dans les traitements informatiques Mém. 1 1979, p. 582*).
   United Kingdom: (*Data Protection Act 1984, 12 July 1984, Her Majesty's Stationery Office, C. 30 9/84 CCP*).

24. Model Provisions on the Protection of Computer Software, World Intellectual Property Organization, Genève, 1978.

25. A. VAN MENSEL, "De bescherming van fabricagegeheimen of technische know-how naar Belgisch recht", *R.W.*, 1981-82 kol. 2001, op kol. 2009.

26. A. LUCAS, *"La protection des créations industrielles abstraites"*, Citée (Paris) 1978, p. 230, no. 358.

27. E. WEISS, "Telecommunications policy: the users need for telecommunication systems. A review of trends in Europe", in *Communication regulation and international business*, Proceedings of a Workshop held at the I.M.I., April 1983, J.F. RADA, G.R. PIPE, ed., 1984, p. 136.
E.g. what is the effect of the privatisation of B.T. ? M.E. BEESLEY, "Liberalisation of the use of British Telecommunications Network", Department of Industry, *Report to the Secretary of State*, London, 1981.
France: Loi, no. 84, 939 du 23 octobre 1984, J.O., 25 octobre 1984, p. 3335.

28. This is the case e.g. in Belgium and apparently also in Japan. Only at academic research level initiatives are under way to adapt the legislation.

29. A solution favoured by countries such as Austria, Finland, Denmark, or Greece.

30. E.g. the U.S.A. Counterfeit Access Device and Computer Fraud and Abuse Act (1984), The Swedish Data Act of 1973.

31. E.g. the Danish Penal Code Amendment Act of June 6, 1985; the Canadian Criminal Law Amendment Act of 1985; the French law of 1988.

32. Revisions are under way in o.a. Belgium, Switzerland, Finland, Norway.

33. OECD – Information Computer Communications Policy, vol. 10. *Computer – related crime: analysis of legal policy*, Paris (1986), 71 p.
COE – *Computer related crime* (final report of the European Committee on Crime Problems, Strasbourg, 1990, 124 p.

34. E.g. U.K. Computer Misuse Act, August 1990.
FRANCE: Loi no. 88-19 du 5 janvier 1988 relative á la fraude informatique, *J.O.*, 6 janvier 1988, p. 231.

35. J. DE HOUWER, "Privacy en grensoverschrijdend dataverkeer", in: Soft- en Hard, ware het niet om de fraude, *IUS*, nr. 7 (1985) p. 92.
DENMARK: *The Danish Private Registers etc. Act No. 293*, 8 June, 1978. *The Danish Public Authorities' Registers Act No. 294*, 8 June, 1978.
FRANCE: *Loi relative á l'informatique, aux fichiers et aux libertés*, No. 78-17, 6 January, 1978, *J.O.*, 7 January, 1978.
ISRAEL: *Protection of Privacy Law*, 5741 - 1981 of 23 February, 1981, *Sefer Ha-Chukkim* No. 1011 of the 5th Adar Bet, 5741 (AA March 1981).
ICELAND: *Act concerning the systematic recording of personal data*, Law nr. 63 of 25 May 25, 1981.
LUXEMBOURG: *Loi du 31 mars 1979 réglementant l'utilisation des données nominatives dans les traitements informatiques*, *Mém.* A 1979, p. 582.
NORWAY: *Act relating to personal data registers etc.*, Law nr. 48 of 9 June, 1978.
AUSTRIA: *Datenschutzgesetz D.S.G.*, October 18, 1978, *Bundesgesetzblatt* 1978, November 28, 1978, 3619–3631.
U.K.: *Data Protection Act 1984*, chapter 35, July 12, 1984, Her Majesty's Stationary Office, C30 9/84 CCP.
WEST GERMANY: *Bundesdatenschutzgesetz*, January 27, 1977, BGBI, I S.201.
SWEDEN: *Data Act* (1973: 289).

36. Even though only a proposal, the Federal Computer System Act proposal (1984) acted as an example for some state laws, e.g. in California (January 1, 1985) Virginia Computer Crimes Act, Senate Bill no. 347, March 7, 1984.

37. OECD, DSTI/ICCP/84.22 (1985) p. 53.

38. D. REMER, *"Legal Care for your software"*, Gower, ershot 1984, 272 p.
W. HANNEMAN, *"The Patentability of Computer Software"*, Kluwer, Deventer, 1985, 258 p.
N. FREED, "Software protection: Introductory observations on the study spon-

sored by the National Commission on new technological uses of Copyright Works",
Jurimetrice Journal, 1977-1978, p. 352.

39. Council of Europe, Computer-related crime, Strasbourg, 1990, Appendix I, No. Ia
and b.

40. B. SPRUYT (B. DE SCHUTTER, collaborating), *"Grensoverschrijdende
Informatica-criminaliteit en de Europese strafrechtelijke samenwerking"*, Kluwer
(Antwerpen), 1989, 163 p.

Technical Approaches
to Thwart Computer Fraud

Joos Vandewalle, René Govaerts, Bart Preneel *

ESAT-COSIC Laboratory,
Department of Electrical Engineering
Katholieke Universiteit Leuven
K. Mercierlaan 94, B-3001 Leuven, Belgium

joos.vandewalle@esat.kuleuven.ac.be

Abstract. In our modern society the data protection and computer security needs can often best be met by classical and/or public key cryptographic techniques. The status and the methods of cryptographic research on algorithms, hardware and software is described, emphasizing on DES and RSA public key. The importance of the cryptographic protection of the future Integrated Broadband Communication Network (IBCN) is overviewed. The evolution of IT security techniques methods and standardization is discussed.

1 Need for Data Security

Computer crimes and information theft have become a serious problem [12, 15, 35]. The general public is quite aware of it. The newspapers and TV stations report about it under a number of fancy names: wiretapping, computer hacking, trapdoor, salamis, data diddling, leakage, masquerading, Trojan horse, virus, worm, bacterium, time bomb, inference, superzapping, scavenging, spoofing, impersonation, piggybacking, software piracy, ... The recent spread of computer viruses has been quite alarming. Although it is difficult to estimate the losses it entails, the technical feasibility of many attacks is without any doubt [12, 15, 35]. Recently computer fraud has been prosecuted in some European countries with the traditional laws and offenders have been sentenced to jail. However it is far better to make computer crimes infeasible by technical means. Cryptography is the science of techniques which make data unintelligible and unmodifiable by outsiders (without detection) and still accessible or verifiable by the legitimate receiver. It has been called by a leading expert in computer security, D. Parker [15] *"The premier safeguard against computer crime"*. Indeed although it is not so easy to implement, cryptography is always an essential element in a comprehensive protection against all these attacks.

---

* NFWO aspirant navorser, sponsored by the National Fund for Scientific Research (Belgium).

2 State of the Art in Cryptography

The basic concept of cryptography (see Fig. 1) is to use a key in order to convert or encrypt the cleartext into the ciphertext so that the intended receiver can with his key and the algorithm obtain the cleartext. It should be difficult for the eavesdropper who has no access to the key of the receiver to obtain the cleartext from the ciphertext. This privacy protection should not be dependent on the fact that the algorithm is secret or public (Kerckhoffs' assumption) in the same way as the safety of a mechanical lock is not dependent on the mechanism but on the key.

In the symmetric or traditional cryptography the keys 1 and 2 of sender and receiver are secret and are the same. Hence it is clear that in a network of about 1000 users one has the burden of the secure exchange of about half a million keys. In the public key or asymmetric algorithms key 1 is different from key 2 and the intended receiver can make his key 1 public for all those who want to send him messages and keep his key 2 secret in order to decrypt his messages. Hence privacy protection of a network with 1000 users only requires the authenticated transmission of 1000 keys of type 1.

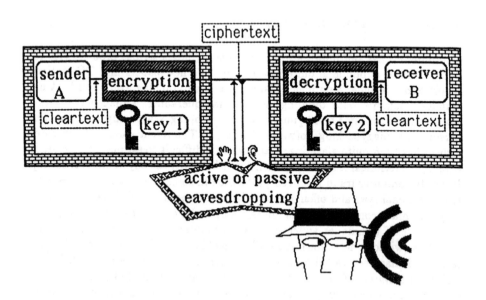

Fig. 1. The basic concept of cryptography

Until 1970 cryptography was mainly pushed by the military and diplomatic needs. The cryptographic techniques used in these domains are not suited for the actual intense and widespread commercial applications. Indeed, either the algorithms are secret, which impedes standardization, or the techniques are too bulky and involved. Since 1970 gradually a different set of algorithms, chips, and electronic equipment has been developed which are called **commercial cryptography**. Only in recent years there has been an intensive use of these techniques and the market still needs more. It is estimated that this market grows by 20% each year.

An important element in commercial cryptography is that public evaluation and reporting is a vital part of the mechanism (see Fig. 2). Each year at least two international conferences are organized on cryptography (Crypto in the US, Eurocrypt in Europe, and recently also Auscrypt in Australia and Asiacrypt in Asia), and several others are involved with data and computer security (Carnahan, ESORICS, National Computer Security Conference, IEEE symposium on Security and Privacy, Securicom ...). Moreover since 1988 a journal, called the Journal of Cryptology, is published. A nice survey of the state of the art on cryptography was published in the May 1988 issue of IEEE proceedings [24]. It was extended an updated into a book [25] in 1991. The fact that many algorithms have been broken in the past few years emphasizes the need for the public evaluation. Although there are currently quite a few algorithms at the initial phases of evaluation, only DES and RSA have reached the mature phases [7, 8, 26], where standardization and commercialization is reached or still continuing. Research in cryptography deals with new methods, protocols, algorithms, and systems. Recently much attention was devoted to zero knowledge protocols, electronic cash, signatures, pseudo-randomness, number theory, smart cards, networks ....

Concerning the security of DES there is a rather general consensus [13] among the cryptographic researchers that it is an extremely good algorithm with an unfortunately small key of 56 bits. Hence it is best used in a multi-encryption scheme (triple encryption with two keys) and with feedback modes instead of the electronic code book mode (ECB) in order to avoid exhaustive key searches and cryptographic attacks. This algorithm is in very wide use, especially in banking applications and has been accepted in a number of standards [1]. It has been reaffirmed by NBS for 5 years in 1987 [26]. The use of DES has been specified in ANSI and ISO/TC68 banking standards, but because of political reasons it is very unlikely that DES will be standardized in ISO/IEC SC27; within this committee it has been decided to register algorithms for confidentiality protection, rather than to standardize them [17]. The register will not guarantee the quality of the algorithm, and the entries in the register might be secret. The security of RSA is also considered to be very good. Today it requires about 1 day for a supercomputer to factor a number of 90 decimal digits with the best methods. It would take about 100,000 years to factor a number of 200 decimal digits. The factorization record (for products of two large primes) is now at 117 decimal digits. Hence keys of 300 decimal digits or more are considered to be secure for the next 5 to 10 years. However the speed of the encryption and

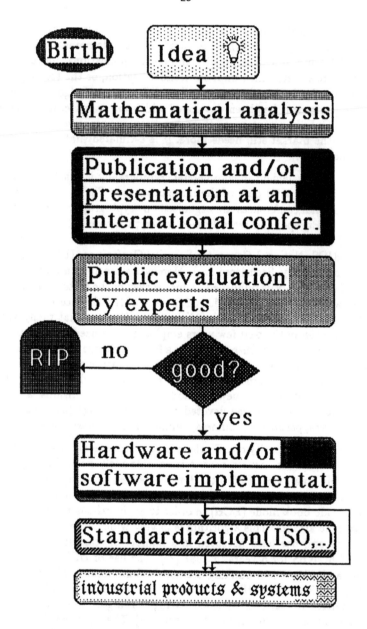

Fig. 2. The life of a commercial cryptographic algorithm

decryption is often too slow. Then the solution is a hybrid system, i.e., one can use RSA for key exchange and afterwards use a fast algorithm like DES for information exchange. For another public key algorithm, based on the discrete logarithm problem, one claims [4] to be able to achieve faster rates at the same security level; during the last years the elliptic curve based algorithms received wide attention. The knapsack algorithm on the other hand is very fast and is very easy to implement. However almost all research has been discontinued, because most of the versions have been broken [7]. The B-Crypt of British Telecom is another interesting algorithm. However it is not clear whether the fact that this algorithm is secret, will limit its use.

Moreover chipcards are very convenient and important in cryptographic applications and electronic payment (electronic checkbook, electronic purse, tele-banking, access control ...). Extensive experiments have been performed in France (Lyon, Blois, and Caen, 1979-1984), Norway, Italy. It is now used in the Belgian TRASEC system for the electronic transmission of financial transactions between the customers and the banks. Many countries and companies have extensive plans for using the chip card (Japan, Mastercard, VISA, OSIS/Teletrust). For further details the reader is referred to the conferences on chipcards (Smart card 2000 Vienna 1987, Amsterdam 1989).

Several standards have been very useful [1, 10] like ANSI X9.9-1982 (revised in 1986), "Financial Institution Message Authentication" [2] and ANSI X9.17-1985, "Financial Institution Key Management (Wholesale)" [3]. Many other standardization efforts are underway (CCITT, OSI, ECMA ...) [17].

In addition it is worth mentioning important developments in the US. Under the commercial COMSEC (Communications Security) Endorsement Program (CCEP) offered by the National Security Agency (NSA) some selected US companies are developing highly secure encryption products [1, 12, 23, 26] based on secret algorithms. Also Secure Telephone Units (STU) are marketed in the US [12, 1] at reasonable prices (around at 2000 $ a piece). The Kerberos authentication system [5] is becoming very popular to protect open networks (it was incorporated into OSF's DCE).

More recently the US government has proposed the digital signature standard (DSS) and the corresponding secure hash standard (SHS) for the electronic verification of the integrity and the source of unclassified information [18, 35]. In addition the "clipper chip" has been put forward for concealment within the US. It contains the secret "skipjack" algorithm, that is a 64-bit block cipher with an 80-bit key. The chip has the specific property that it reveals the secret key encrypted under a master key. This master key is stored in a secure place (divided over two or more escrow agencies), and can be obtained with a warrant that authorizes a line tap. The skipjack is a secret algorithm and hence cannot be subject to a public scrutiny which opens a debate about their strength and the existence of trapdoors. Moreover this chip can only be produced within the US; it might be allowed to use it outside the US.

Industrial software products for computers and network security that often have been evaluated by the American National Computer Security Center are SCOMP (Honeywell), RACF (IBM), ACF2 (SKK), Panoramics, Top Secret (CA), ...[1].

It is often argued at the management level that security is an expensive way to degrade a computer system. However the professional attitude is that security is a management tool which is effective in controlling the use, i.e., the availability, the integrity, and the confidentiality of data. A global security policy should be devised including the culture, the standards, and the procedures, with specific regulations on hardware, software, communications, environment, personnel, contingency, and disaster recovery.

3 The Worldwide Evolution of IT Security

It is now generally and worldwide acknowledged that security is an essential component in Information Technology (IT) and that IT plays an important role in almost all sectors of societies like those in the US, Japan, and Europe.

In this context, IT security means [31]:

confidentiality: information is only disclosed to those users who are authorised to have access to it;

integrity: information is modified only by those users who have the right to do so;

availability: information and other IT resources can be accessed by authorised users when needed.

Every IT system or product will have its own requirements for maintenance of confidentiality, integrity, and availability. In order to meet these requirements it will contain a number of **security functions**, covering, for example, areas such as access control, auditing, and error recovery. Appropriate confidence in these functions will be needed: in [31] this is referred to as **assurance**, whether it is confidence in the **correctness** of the security functions (both from the development and the operational points of view) or confidence in the **effectiveness** of those security functions.

In response to these needs and requirements of the users a document called Trusted Computer System Evaluation Criteria (TSEC) [30], commonly know as the "Orange Book", has been published by the US government. This was followed by many other countries especially in Europe. The ITSEC document [31] has been drafted in order to harmonize and extend the criteria of France, Germany, the Netherlands, and the United Kingdom. The corresponding ITSEM document [32] specifies the evaluation procedure.

When considered from the consumer side, there is a vast increase in personal computation resources in a broad spectrum of products. IC technology and VLSI will make these products even more performant and cheaper. Many of these products require on line transactions. Moreover the services are often mobile (the cellular radio, the personal communicator). From all this, one can easily infer an increase in the threats both in the strength and in the frequency. For example, mobile services can be easily intercepted.

Hence the market for IT security is expected to grow considerably in the coming years: trusted computer programs will be widespread and cryptography will be used on a wide scale in trusted computer systems. G. Simmons has put it very pointedly in one of his predictions for the nineties: *"By the end of the decade (that sounds more modest than "the turn of the century"), virtually everyone will have and will routinely use a cryptographic-based identity and transaction identifier for everything from ATM's, point-of-scale transactions, access control (to the phone system, data banks, etc.) to voting. I don't think the identifier is apt to be a calculator-like object, but much more apt to be something one keeps and wears all the time: the 2001 version of the dog-tags I wore in the Army for nearly five years – with similar, but low-tech function. The key words in this prediction are "virtually everyone" and "routinely": there is no uncertainty in predicting "many" and "for some purposes" or "occasionally". "*

A Japanese study [34] estimates that the Japanese market is between one tenth and one twentieth of that of the US. It is expected to grow each year by about 15% from now until 2010. In the 21st century the security industry should have established an invincible position and users should have firmly established security policies. This includes several relevant security industry segments like access control, cryptography, backup services [34].

On the other hand we should be careful not to howl with the wolves about computer crime. D. Parker [16] has recently debunked 17 myths that circulate among IT people and journalists. His main statements are the following: There are no valid figures for computer crime losses. It is not proven that computer crime is mostly caused by insiders, or is motivated by greed. Computer viruses are not a major threat. There are a great variety of mechanisms for information loss. Information in computers is not more vulnerable than spoken, printed or displayed information. Eavesdropping and tapping data wires is not such a common criminal activity. Although the incidence of business crime is diminishing with increasing use of computers, the loss per case is increasing. The primary business confidentiality policy should be based on the need-to-withhold rather than on the need-to-know. Automatic dial back is not the best telephone access control. Authentications of identity should be based on very diverse items. Tiger team testing is dangerous, ineffective, inefficient, and unethical in a non-military environment. In addition to good controls and practices, people are essential in order to achieve effective information security. Quantitative risk assessment is unreliable. Increasing computer usage is improving personal privacy not destroying it.

These statements should not be misinterpreted as if computer security is of no concern at all. In fact [35] computer viruses and network attacks are becoming more and more sophisticated and widespread. Both the number of viruses and the number of incidents keeps increasing but not exponentially as some researchers have predicted.

4 Evolution of IT Security in Europe

Most of the observations made in the previous section are of course also valid for Europe. In addition, there are a number of specific evolutions, rapid changes, and attitudes.

First of all, there is a unified European market, which has now been combined with the opening of many East European countries. This new configuration opens unprecedented markets, and opportunities to cooperate and interact much more intensely. When this is combined with the generally good education in Eastern Europe, an intense economic growth is expected. However this euphoria has calmed down to some realism.

By the end of 1992 the European Community has set up a unified European market of about 300 million customers. In view of this market integrated broadband communication (IBC) using fiber optics is planned for commercial use in 1995. This IBC will provide high speed channels (64 kbps, 2 Mbps and more) of image, voice, sound and data communications and will support a broad spectrum of services like telex, telefax, telephony, teletex, videotex, electronic mail, telenewspaper, teleconferencing, videoconferencing, cable TV, telebanking, teleshopping, home banking, EFT, POS, mobile telephony, paging, alarm service, directory services, etc. These services can be home based, office based, (private or public) manufacturing or mobile. They may include dialogue service or messaging or retrieval or a distribution service. It is clear that the majority of these services offered in future networks are crucially dependent on security.

Also for the IT Security several European countries felt the need to harmonize the criteria [31] . This is certainly an important document both for the users and the manufacturers. When products and systems are evaluated against ITSEC criteria, system managers and system integrators have a quantifiable level of assurance for the selection of products and systems. However as explained in detail in the contribution of D.W. Roberts [22], ITSEC has some limitations. It does not deal with tamper resistant devices, and it does not describe accurately the effectiveness of the data security. Moreover the delay of certification implies that the computers are often outdated by the time they are certified. Last but not least the ITSEC criteria do not eliminate completely the need for trust of the consumer in the manufacturer. Let us also mention here that Europe has important initiatives at the network security level like the RACE project SESAME.

5 IT Security Issues and Conclusions

It is likely that the European market will harmonize and expand. Hence the European IT security criteria are likely to spread and extend. Hence one should expect on the market in the next five years an explosion of devices, systems, tokens, and other security products based on cryptography. One of the key issues here is to make a trade-off between three often conflicting demands: security, speed, and user-friendliness. Another problem is related to the procedure one should adopt when a widely used algorithm like DES would be broken. Although some progress is made, there are also still important legislative obstacles between the European Countries which may impede international EFT. For example one country may allow EFT encryption, while others may only allow authentication [1, 20].

Although the European market will unify and even a common European currency is expected, Europe will still be marked by many different cultures and languages. How can information technology and its security be efficiently and effectively worked out so that a free flow of people and products can be supported ? Moreover since many services (like bank services) move closer to the users, how can IT support multilingual and multicultural services ?

Most companies have products from several vendors. Hence it is not so obvious to implement a company wide security policy and to define the responsibilities. In addition it may be difficult to apply these principles in court. Here the differences between the legal systems in the different European countries may be a major bottleneck. For example in Belgium the hackers that entered the BISTEL government computer network were condemned because they were stealing electric energy, which is a rather weak argument from a technical point of view.

Since cryptography is an essential element in a global security concept, what are the future developments we can expect in this area? (public key, zero knowledge protocols, anonymous payment, ...)

It is hoped that the mutual strengths and needs of the users and manufacturers can be combined in order to provide secure systems for information technology in Europe. In the coming years, Europe is expected to move quickly and both the users and the manufacturers can greatly benefit from these opportunities if good communication is established and efforts are combined.

References

1. M.D. Abrams and H.D. Powell, *"Tutorial computer and network security,"* IEEE Computer Society Press, Los Angeles, 1987.

2. *"American National Standard for Financial Institution Message Authentication (Wholesale),"* X9.9-1986 (Revised), ANSI, New York.

3. *"American National Standard for Financial Institution Key Management (Wholesale),"* X9.17-1985, ANSI, New York.

4. I. Blake, P. van Oorschot, and S. Vanstone, "Complexity issues for public key cryptography," in *"Performance Limits in Communication Theory and Practice,"* J.K. Skwirzynksi, Ed., Kluwer, 1988, pp. 75–97.

5. B. De Decker, "Unix security & Kerberos," this volume.

6. D. Denning, "The clipper chip: a technical summary," April 21, 1993.

7. Y. Desmedt, "What happened with knapsack cryptographic schemes?" in *"Performance Limits in Communication Theory and Practice,"* J.K. Skwirzynski, Ed., Kluwer, 1988, pp. 113–134.

8. W. Diffie, "The first ten years of public key cryptology," in *"Contemporary Cryptology: The Science of Information Integrity,"* G.J. Simmons, Ed., IEEE Press, 1991, pp. 135–175.

9. W. Fumy and H.P. Rieß, *"Kryptographie. Entwurf und Analyse symmetrischer Kryptosysteme,"* Oldenburg, München, 1988.

10. B. Greenlee, "Requirements for key management protocols in the wholesale financial service industry," *IEEE Communications Magazine*, Sept. 1985, pp. 22–28.

11. F. Hoornaert, M. Decroos, J. Vandewalle, and R. Govaerts, "Fast RSA-hardware: dream or reality ?," *Advances in Cryptology, Proc. Eurocrypt'88, LNCS 330*, C.G. Günther, Ed., Springer-Verlag, 1988, pp. 257–264.

12. J. Horgan, "Thwarting information thieves," *IEEE Spectrum*, July 1985, pp. 30–41.

13. J.L. Massey, "An introduction to contemporary cryptology," in *"Contemporary Cryptology: The Science of Information Integrity,"* G.J. Simmons, Ed., IEEE Press, 1991, pp. 3–39.

14. D. Newman and R. Pickholtz, "Cryptography in the private sector," *IEEE Communications Magazine*, Vol. 24, No. 8, August 1986, pp. 7–10.

15. D. Parker, *"Fighting computer crime,"* Ch. Scribner's Sons, New York, 1983.

16. D. Parker, "Seventeen information security myths debunked," *Proc. IFIP Workshop Finland*, June 1990.

17. B. Preneel, "Standardization of cryptographic techniques," this volume.

18. B. Preneel, R. Govaerts, and J. Vandewalle, "Information authentication: hash functions and digital signatures," this volume.

19. K. Rihaczek, "Datensicherheit amerikanisch," *Datenschutz und Datensicherung*, 1987, pp. 240–245.

20. Rihaczek K., "Ein Kompromiszvorschlag zur Datenverschüsselung," *Datenschutz und Datensicherung*, 1987, pp. 299–303.

21. R.L. Rivest, A. Shamir, and L. Adleman, "A method for obtaining digital signatures and public-key cryptosystems," *Communications ACM*, Vol. 21, February 1978, pp. 120–126.

22. D.W. Roberts, "Evaluation criteria for IT security," this volume.

23. R. Rosenberg, "Slamming the door on data thieves; Can the NSA create and enforce a new encryption standard ?" *Electronics*, Feb. 3, 1986, pp. 27-31.

24. G.J. Simmons, Ed., "Special section on cryptology," pp. 515–518, pp. 533–627, *Proc. IEEE*, Vol. 6, No. 5, May 1988. (contains early versions of [8, 13, 26])

25. G.J. Simmons, Ed., *"Contemporary Cryptology. The Science of Information Integrity,"* IEEE Press, 1991.

26. M.E. Smid and D.K. Branstad, "The Data Encryption Standard: past and future," in *"Contemporary Cryptology: The Science of Information Integrity,"* G.J. Simmons, Ed., IEEE Press, 1991, pp. 43–64.

27. J. Vandewalle, R. Govaerts, W. De Becker, M. Decroos, and G. Speybrouck, "Implementation study of public key cryptographic protection in an existing electronic mail and document handling system," *Advances in Cryptology, Proc. Eurocrypt'85, LNCS 219*, F. Pilcher, Ed., Springer-Verlag, 1986, pp. 43–49.

28. J. Vandewalle and R. Govaerts, "Trends in data security," *Proc. Secubank '88, Secure Banking Information Systems*, Datakontext-Verlag, Köln, 1988, pp. 32–38.

29. I. Verbauwhede, F. Hoornaert, J. Vandewalle, H. De Man, and R. Govaerts, "Security considerations in the design and implementation of a new DES chip," *Advances in Cryptology, Proc. Eurocrypt'87, LNCS 304*, D. Chaum and W.L. Price, Eds., Springer-Verlag, 1988, pp. 287–300.

30. *"Trusted Computer Systems Evaluation Criteria,"* DOD 5200.28 - STD, Department of Defense, United States of America, December 1985. (Orange book)

31. *"Information Technology Security Evaluation Criteria (ITSEC),"* Provisional Harmonised Criteria, Version 1.2, June 1991, ISBN 92-826-3004-8, Catalogue number CD-71-91-502-EN-C.

32. *"Information Technology Security Evaluation Manual (ITSEM),"* Version 0.2, April 1992.

33. *"OTR 200, Race Workplan,"* Commission of the European Communities, 1988, Rue de la Loi 200, B-1049 Brussels, Belgium.

34. "The Security Industry in 2010," *Japan Computer Quarterly*, 1990, pp. 3-17.

35. "Special report on data security," *IEEE Spectrum*, Vol. 29, No. 9, August 1992, pp. 18-44.

Section 2

Theory

Public key cryptography

Marijke De Soete

Philips Industrial and Telecommunication Systems
Tweestationsstraat 80, B–1070 Brussels

Abstract. This paper deals with public key cryptosystems and some of their applications such as password encryption and digital signatures. The necessary mathematical background is also provided.

1 Introduction

Cryptology is the general term comprising *Cryptography*, i.e., the art of providing secure communications over insecure channels, and *Cryptanalysis*, i.e., the dual art of breaking into such communication systems. In this paper the focus is on the "new" type of cryptography, known as *public key cryptography*, which was introduced in the late seventies. We deal with cryptosystems based on it and some of their applications. We also include discussions on some security provisions to counter cryptanalysis. Public key cryptography requires a considerable mathematical background. We therefore start with a section on some basic notions in number theory which are necessary to understand the following sections.

2 Mathematical Background

2.1 Modular Arithmetic

Definition. For integers a, b and n, $n \neq 0$, we define "a *is congruent to b modulo n*" if and only if

$$a - b = k \cdot n, \text{ i.e. } n \mid a - b.$$

We use the notation $a \equiv b \bmod n$ and the integer n is called the *modulus* of the congruence.

Example. $17 \equiv 7 \bmod 5$ since $17 - 7 = 2 \cdot 5$.

The following properties are easily proved directly from the definition:

- $a \equiv a \bmod n$
- $a \equiv b \bmod n \Leftrightarrow b \equiv a \bmod n$
- $a \equiv b \bmod n$ and $b \equiv c \bmod n \Rightarrow a \equiv c \bmod n$.

This means that for fixed n, "congruence modulo n" is an *equivalence relation*. Each equivalence class (*residue class*) contains exactly one element in $[0, n-1]$ by which it can be represented.

Example. $17 \equiv 2 \bmod 5$, hence $17 \bmod 5$ is represented by 2.

Theorem 1. *For integers a_1, a_2 and n, $n \neq 0$, there holds:*

$$(a_1 \; op \; a_2) \bmod n \equiv [a_1 \bmod n \; op \; a_2 \bmod n] \bmod n \, ,$$

where "op" stands for "+" or ".".

It follows that the set of equivalence classes for a fixed modulus n is a *commutative ring*.

Theorem 2. *For integers e, n and t, with $n \neq 0$ and $0 < t \leq n-1$, there holds:*

$$e^t \bmod n \equiv [\prod_{i=1}^{t}(e \bmod n)] \bmod n \, .$$

This property is very important for the implementation of modular exponentiations.

Unlike integer arithmetic, modular arithmetic allows the calculation of inverses in some cases. This means that for some $a \in [0, n-1]$ an integer $x \in [0, n-1]$ can be found such that

$$a \cdot x \equiv 1 \bmod n \, .$$

Example. $3 \cdot 7 \bmod 10 = 21 \bmod 10 \equiv 1$, hence 3 and 7 are inverses for the congruence modulo 10.

The next theorem shows which integers have an inverse in modular arithmetic.

Theorem 3. *An integer $a \in [0, n-1]$ has a unique inverse modulo n if and only if a and n are relatively prime, i.e., $\gcd(a, n) = 1$ (where $\gcd$ denotes the greatest common divisor).*

Inverses can be calculated by means of the Euclidean algorithm.

Example. Find $160^{-1} \bmod 841$, i.e., find the inverse of 160 modulo 841.

Since $\gcd(841, 160) = 1$, we obtain by the Euclidean algorithm:

$$841 = 5 \cdot 160 + 41$$
$$160 = 3 \cdot 41 + 37$$
$$41 = 1 \cdot 37 + 4$$
$$37 = 9 \cdot 4 + 1.$$

Hence

$$\begin{aligned}
1 &= 37 - 9 \cdot 4 = 37 - 9 \cdot (41 - 1 \cdot 37) \\
&= 10 \cdot 37 - 9 \cdot 41 = 10 \cdot (160 - 3 \cdot 41) - 9 \cdot 41 \\
&= 10 \cdot 160 - 39 \cdot 41 = 10 \cdot 160 - 39 \cdot (841 - 5 \cdot 160) \\
&= 205 \cdot 160 - 39 \cdot 841.
\end{aligned}$$

It follows that $160^{-1} \bmod 841 \equiv 205$.

Theorem 4 (Fermat's theorem). *Let p be a prime. Every integer a satisfies:*

$$a^p \equiv a \bmod p$$

and every integer a not divisible by p satisfies:

$$a^{p-1} \equiv 1 \bmod p.$$

Example. For $p = 11$, $a = 2$, calculate $2^{10} \bmod 11$:

$$2^4 \bmod 11 \equiv 16 \bmod 11 \equiv 5$$
$$2^8 \bmod 11 \equiv 5^2 \bmod 11 \equiv 25 \bmod 11 \equiv 3$$
$$2^{10} \bmod 11 \equiv 3 \cdot 2^2 \bmod 11 \equiv 12 \bmod 11 \equiv 1.$$

Note that Fermat's Theorem can be used to check whether a number is composite:

If $\gcd(a, n) = 1$ and $a^{n-1} \not\equiv 1 \bmod n$ then n is composite.

However, it can not be used as a primality test. For example, $\gcd(2, 341) = 1$ and $2^{340} \equiv 1 \bmod 341$ but $341 = 11 \cdot 31$.

Definition. For an integer n, the *Euler totient function* $\Phi(n)$ denotes the number of positive integers less than n which are relatively prime to n.

It follows immediately:

- for p prime: $\Phi(p) = p - 1$
- for $n = p \cdot q$, with p and q prime: $\Phi(n) = \Phi(p) \cdot \Phi(q) = (p - 1) \cdot (q - 1)$

Example. For $p = 3$, $q = 5$ we obtain:

$$\Phi(15) = (3 - 1) \cdot (5 - 1) = 2 \cdot 4 = 8 \,.$$

Indeed, the set of integers less than 15 and relatively prime to 15 is equal to $\{1, 2, 4, 7, 8, 11, 13, 14\}$.

Theorem 5 (Euler's generalization). *For all integers a, n with $\gcd(a, n) = 1$ there holds:*
$$a^{\Phi(n)} \equiv 1 \bmod n \,.$$

It follows from this theorem that to solve $a \cdot x \equiv 1 \bmod n$, with $\gcd(a, n) = 1$, we have to calculate

$$x \equiv a^{\Phi(n) - 1} \bmod n \,.$$

Example. For $a = 3$ and $n = 7$ we have $\Phi(7) = 7 - 1 = 6$.
Hence we obtain for the inverse x of a

$$x \equiv a^{\Phi(n) - 1} \bmod n \equiv 3^5 \bmod 7 \equiv 5 \,.$$

Indeed, $3 \cdot 5 \bmod 7 \equiv 1$.

Hence there are two different means by which the modular inverse of an integer (if it exists) can be calculated:

1. using $\Phi(n)$ if the value is known
2. applying the extension of Euclid's algorithm for the gcd.

Theorem 6 (Chinese remainder theorem). *For a system of congruences*

$$x \equiv a_1 \bmod m_1$$

$$x \equiv a_2 \bmod m_2$$

$$\vdots$$

$$x \equiv a_r \bmod m_r \,,$$

with $\gcd(m_i, m_j) = 1$, $i \neq j$, there exists a unique solution modulo $M = m_1 \cdot m_2 \cdot \ldots \cdot m_r$. This solution is given by $x = \sum_i a_i \cdot M_i \cdot N_i$ with $M_i = M/m_i$ and $N_i \equiv M_i^{-1} \bmod m_i$.

Example. To solve the equation $7 \cdot x \bmod 15 \equiv 1$ we apply the Chinese remainder theorem. Since $15 = 3 \cdot 5$ we first solve:

$$7 \cdot x_1 \equiv 1 \bmod 3$$

$$7 \cdot x_2 \equiv 1 \bmod 5 \,,$$

giving $x_1 = 1$ and $x_2 = 3$. So we have to find a common solution to

$$x \equiv 1 \bmod 3$$

$$x \equiv 3 \bmod 5$$

by calculating

$$x = (\frac{15}{3} \cdot x_1 \cdot N_1 + \frac{15}{5} \cdot x_2 \cdot N_2) \bmod 15,$$

where N_1, N_2 are the solutions of

$$\frac{15}{3} \cdot N_1 \equiv 1 \bmod 3$$

$$\frac{15}{5} \cdot N_2 \equiv 1 \bmod 5.$$

This results in $N_1 = N_2 = 2$ and hence

$$x = (1 \cdot 5 \cdot 2 + 3 \cdot 3 \cdot 2) \bmod 15 \equiv 28 \bmod 15 \equiv 13.$$

Indeed, 13 is the inverse of 7 mod 15.

2.2 Primes

Prime numbers play a very important role in public key cryptography. The two major functions are generating a prime and testing a number for primality. Generating a prime means finding a prime within a given range $[r_1, r_2]$. Testing a number for primality means testing whether a given number is prime or not.

Before discussing primality tests we have to introduce the notion of a *"pseudoprime"*.

Definition. We call an integer s such that

$$a^{s-1} \equiv 1 \bmod s, \text{ for some } a, \text{ with } 1 < a < s$$

a *pseudoprime to the base a*.

Theorem 7 (Fermat). *If s is prime and $\gcd(a, s) = 1$, then s is a pseudoprime to the base a.*

Definition. Let s be an odd integer such that

$$s - 1 = 2^v \cdot s', \quad s' \text{odd}.$$

Then s is a *strong base pseudo prime to the base a* if either

$$a^{s'} \equiv 1 \bmod s$$

or

$$a^{2^k \cdot s'} \equiv -1 \bmod s \,,$$

for some k, $0 \le k \le v$.

Compositeness test. If, for a given base a, $1 < a < s$, s is not a strong pseudoprime to the base a, then s is composite.
If s is composite, at least $3 \cdot (s-1)/4$ bases for which s is a pseudoprime exist.

Primality test of Rabin. If, for k random integers a_i, $1 < a_i < s$, s is a strong pseudoprime to the base a_i, then s is prime with an error of $1/4^k$.

2.3   One-Way Functions

Definition. A function

$$f : \begin{array}{cc} X & Y \\ x & \mapsto f(x) = y \end{array}$$

is a *one-way function* $\Leftrightarrow$

 - $f(x)$ is easy to compute for all x in X,
 - given y in X, it is computationally infeasible to find an x in X satisfying $f(x) = y$.

The existence of one-way functions is still an open problem. Candidates are those functions which we know to compute efficiently and for which no efficient algorithms are known which "invert" them.

A very simple example of a candidate for a one-way function is *integer multiplication* for large integers. Indeed, it is very easy to multiply two large integers whereas the most powerful computers with the best known algorithms cannot factor (in reasonable time) a number of say two hundred digits which is the product of two primes of approximately equal size.

Another important candidate is *modular exponentiation* with fixed basis and modulus. Given integers a (*the base*) and n with $a \in [1, n-1]$, the modular exponentiation $a^m \bmod n$ can be computed efficiently even if all the parameters are several hundred digits long. As an easy example we mention $a^{25} = (((a^2 \cdot a)^2)^2)^2 \cdot a)$ which shows that only four squarings and two further multiplications are needed. In a computation of $a^m \bmod n$, the reduction modulo n should be done after each squaring and each multiplication in order to avoid calculations with very large integers (see Theorem 2 in Sect. 2.1).
The inverse operation is known as the *discrete logarithm*: given integers a, n and x, find some integer m such that

$$a^m \bmod n \equiv x \,.$$

Example. Since $5^4 \bmod 21 \equiv 16$, 4 is a solution to the discrete logarithm of 16 in base 5 modulo 21. Notice that also 10 is a solution for $5^{10} \bmod 21 \equiv 16$.

On the other hand one can verify that 3 has no logarithms to the base 5 modulo 21, this means that the equation $5^x \bmod 21 \equiv 3$ has no solution.

Although large modular exponentiations can be computed very efficiently, no algorithm is known at present which does the calculation of discrete logarithms in a reasonable time.

One-way functions cannot be used directly in cryptosystems since the receiver must be able to retrieve the message m from a received cryptogram $f(m)$. However, they prove to be useful for password encryption (see Sect. 5.1).

A notion more relevant to public key cryptography is that of the trapdoor one-way functions which are treated in the next section.

2.4 Trapdoor One-Way Functions

Definition. A *trapdoor one-way function* is a one-way function with the property that some additional information, the so called *trapdoor*, makes the function invertible.

A candidate for a trapdoor one-way function is again based on modular exponentiation. For fixed exponent m and fixed modulus n we consider the inverse operation of $a^m \bmod n$. This operation is known as taking *"the m-th root of x modulo n"*, with $x \equiv a^m \bmod n$.

Example. 5 is a 4-th root of 16 modulo 21 since $5^4 \bmod 21 \equiv 16$. You may also check that 2 is a 4-th root of 16 modulo 21.

Whenever m and n are fixed, $a^m \bmod n$ can be calculated efficiently. Contrary to the discrete logarithm problem, however, there exists an efficient algorithm which allows, for any given x, to take m-th roots of $x \bmod n$ or to find out that they are non existing. But it is not known how to construct such an efficient algorithm if only m and n are given. An efficient algorithm for inverting the modular exponentiation for fixed m and n requires the knowledge of the decomposition of n into prime factors. Precisely this information serves as the secret trapdoor. When this trapdoor is known, the Euler Φ-function can be used to calculate the m-th root modulo n as we will see when discussing the RSA cryptosystem (see Sect. 3.4).

3 Public Key Systems

3.1 Introduction

The security of a cryptosystem designed for general use is not based on the secrecy of the algorithm employed but rather on a relatively short secret parameter known as *the key*. An obvious requirement for the security is the existence of a very large number of possible keys. Notice however this is not sufficient to guarantee a safe cryptosystem. Cryptanalysis of some cryptosystems could for instance be based on the variation in natural language letter frequencies which can facilitate considerably the search for the key even for a large key space.

A cryptosystem is a *secret key* system if some secret piece of information (the key) has to be agreed upon ahead of time between any two parties wishing to communicate through the cryptosystem. In such a cryptosystem comprising a group of n users a total of $n \cdot (n - 1)/2$ secret keys are needed if each two of them want to communicate securely under a unique key. This also means that every user has to store $n - 1$ secret keys. It follows immediately that secure key distribution is a non-negligible problem when such a cryptosystem is used in a large network. For the number of keys is growing quadratically with the number of users.

In 1976, Diffie and Hellman overcame this difficulty by introducing the notion of public key cryptography [11]. A similar idea was independently defined by Merkle [17]. The first proposed practical implementation is due to Rivest, Shamir and Adleman in 1978 [23]. Secure communication over insecure channels between two totally unacquainted parties became conceivable. It is important to understand that the security of these systems relies on computational complexity theory and that they are completely insecure according to Shannon's information theory.

The idea behind the notion of public key cryptography is that whoever enciphers a message does not need to be able to decipher it. In such systems each user selects a private *secret key* which is only known to him/her, while he/she distributes the corresponding *public key* to all users wanting to communicate with him/her. The system is only secure if it is infeasible to figure out the secret key from the corresponding public key information. In a system comprised of n users, there is only a total of $2 \cdot n$ keys.

3.2   Public Key Distribution System

As mentioned before, one of the major difficulties with large scale multi-user secret key cryptosystems is that any two users must share a secret key. Assume now that they have not yet established a common secret key and that they want to communicate securely. The conventional solution would be to meet physically in order to exchange a secret key, or to make use of some kind of trusted courier. Both these solutions are slow and expensive and moreover they may not be all that safe. The purpose of a *public key distribution system* is to allow two such users to come up with a secret key as the result of a communication over an insecure channel, in a way that an eavesdropper cannot figure out the key.

The first protocol to achieve this goal was proposed by Diffie and Hellman [11] in 1976. It is based on the discrete logarithm problem introduced in Sect. 2.3. Let n be some large integer and let a be another integer strictly between 1 and $n - 1$. As a first step the two users, say A (Alice) and B (Bob), agree on n and

a over the insecure channel or they could be standard parameters for all users of the system. Then A chooses a large integer x and computes $X = a^x \bmod n$. Similarly, B chooses y and computes $Y = a^y \bmod n$. At this point, A and B exchange X and Y over the insecure channel but they keep x and y secret (only A knows x and only B knows y). Finally, A computes $Y^x \bmod n$; similarly B computes $X^y \bmod n$. Both these values are equal since it amounts to $a^{x \cdot y} \bmod n$ either way. This is the secret key k they wished to establish in common. The protocol is shown in Fig. 1.

$$
\begin{array}{cc}
A & B \\[4pt]
\text{secret } x & \text{secret } y \\[6pt]
X = a^x \bmod n \quad \longrightarrow & \\
\longleftarrow & Y = a^y \bmod n \\
Y^x \bmod n \equiv & X^y \bmod n \equiv \\
(a^y \bmod n)^x \bmod n & (a^x \bmod n)^y \bmod n \\
\equiv a^{y \cdot x} \bmod n \quad = k = & \equiv a^{x \cdot y} \bmod n
\end{array}
$$

Fig. 1. The Diffie-Hellman scheme

An eavesdropper has to figure out k from the information sent over the insecure channel: a, n, X and Y. But figuring out x or y from this information is precisely the discrete logarithm problem (see Sect. 2.3).

The choice of a and n has a substantial impact on the scheme's efficiency and safety. In order to increase the range of potential values for k it is important that the modular exponentiation be as nearly one-to-one as possible. For every prime number n, integers a can always be found such that $a^x \bmod n$ takes on every value in $[1, n-1]$ as x covers the range. Such an integer a is called a *generator for the cyclic group* $GF(n) \setminus \{0\}$. One could also perform all the calculations in the Galois field $GF(2^s)$, a technique which goes beyond the scope of this paper. It allows for efficient computation of the multiplication (and hence the exponentiation) because it avoids the need for computing carries and modular reductions (see [2], [3] and [19]). However, a longer key size is required to achieve the same security level as with a prime modulus [4], [5], [26].

Although very large discrete logarithms may be infeasible to compute, the reader should be warned that there exist algorithms far better than exhaustive search for this task. The best currently known algorithms are described in [8] when n is a prime number and in [20] for computations in $GF(2^s)$. This should be carefully considered when choosing a key size. For instance some attacks can be avoided when using a prime modulus by choosing p such that $p - 1$ has at least one large prime factor.

3.3 Public Key Cryptosystem

A public key distribution system allows two parties to establish a shared piece of secret information. However, neither party has any direct influence on the precise contents of this information. Should A wish to send a specific message, say M, to B, the use of a public key distribution system would have to be followed by that of a secret key cryptosystem in which the initially shared information would serve as secret key.

Public key cryptosystems are used *directly* to encipher messages. Similar to secret key cryptosystems, a public key cryptosystem consists of an enciphering algorithm E and a deciphering algorithm D, but the keys used with this algorithms differ. In fact E used with the public key k_e is trapdoor one-way, this means that it is infeasible to compute the secret key k_d for the deciphering algorithm from knowledge of the description of E and of k_e. Moreover the algorithms are such that

$$D_{k_d}[E_{k_e}(M)] = M \, .$$

The system is shown in Fig. 2 where C denotes the enciphered message $E_{k_e}(M)$.

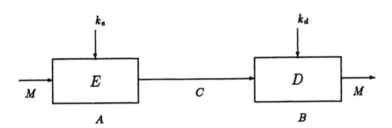

Fig. 2. Public key cryptosystem

In a public key cryptosystem each user selects his/her secret key k_d and his/her public key k_e which he/she makes publicly available, for example through the use of some directory. Whenever a user A wishes to send an enciphered message to user B she looks up B's public key and uses it to encipher the message. Only the genuine receiver B can decipher it by using his secret key.

Contrary to secret key cryptosystems, user A is not better off than an eavesdropper intercepting some ciphertext if she keeps the ciphertext but looses the clear text. Also contrary to secret key cryptosystems, if an eavesdropper intercepts a ciphertext and if he knows the recipient's public key, he can then easily verify any specific guesses as to what the cleartext might be. The cryptanalyst's ability to come up with ciphertexts for cleartext messages of his choice always allows a so called *chosen ciphertext attack*.

Public key cryptosystems rely on the existence of trapdoor one-way functions. Just as was the case with trapdoor one-way functions, we must therefore be satisfied with candidates for public key cryptosystems. The two best known candidates were designed shortly after the introduction of the notion "public key cryptography" by Diffie and Hellman [11]. One of them, Merkle and Hellman's so called *knapsack cryptosystem* [16] was eventually broken [1], [10], [25]. Although some unbroken variations on the original scheme still exist, it is not very advisable to trust them. The most famous public key cryptosystem which we will describe in the next section remains yet undefeated.

3.4   The RSA System

The very first public key cryptosystem ever proposed in the open literature is due to Rivest, Adleman and Shamir in 1978 [23]. It is generally known as the RSA cryptosystem. It is based on modular exponentiation with fixed exponent and modulus (see Sect. 2.4).

Let p and q be two large distinct primes. Let $n = p \cdot q$ and let $\lambda(n) = \text{lcm}(p-1, q-1) = \Phi(n)/\gcd(p-1, q-1)$. Choose e coprime to $\lambda(n)$ and compute $d = e^{-1} \bmod \lambda(n)$. Note that d only can be calculated if the trapdoor, i.e., the primes p and q, are known. Then $k_d = < d, n >$ will be the secret key and $k_e = < e, n >$ the public key for the system.
Enciphering a given message m is given by:

$$C = E_{k_e}(M) = M^e \bmod n,$$

while deciphering the cryptogram C is given by

$$D_{k_d}(C) = D_{k_d}(M^e \bmod n) = (M^e \bmod n)^d \bmod n$$

$$\equiv M^{e \cdot d} \bmod n \equiv M \bmod n,$$

since $e \cdot d \equiv 1 \bmod \lambda(n)$ (see Sect. 2.1).

Example. For the values $p = 19$, $q = 23$ we have $n = 437$. Choosing $e = 13$ one easily verifies that $d = 61$ since $13 \cdot 61 = 793 \equiv 1 \bmod \lambda(437)$.
Enciphering for instance the message $M = 123$ results in $C = 123^{13} \bmod 437 \equiv 386 \bmod 437$.
Deciphering C gives $386^{61} \bmod 437 \equiv 123 \bmod 437 = M$.

The implementation of this cryptosystem requires an efficient algorithm for the modular exponentiation since typically at least 512 bit parameters are used (see [14]). An even more efficient algorithm can be employed for the decryption if p and q are known, based on the Chinese remainder theorem (see Sect. 2.1 and [21]).
For the generation of the two large primes p and q one of the primality tests such as the Miller Rabin test can be applied (see Sect. 2.2). Since the system is broken when knowing the factorization of n, special precautions have to be

taken into account concerning the choice of p and q. In particular, the greatest common divisor of $p - 1$ and $q - 1$ should be small and they both should have large prime factors. For an extensive discussion on these properties we refer to [9].

It should be noted that for the factorization of large integers more efficient methods than exhaustive search do exist. The most famous ones are the quadratic sieve method and the number field sieve method [15].

Despite all its advantages over secret key cryptosystems, RSA is substantially slower than secret key cryptosystems. Even the fastest RSA implementations known nowadays (see [24]) only make the system practical for the exchange of keys used in hybrid systems.

4 Hybrid Systems

Public key cryptography has some important advantages over secret key cryptography when it comes to the distribution of keys. However, when a large amount of information has to be communicated, it may be that the use of RSA would be too slow whereas the use of secret key cryptosystem could be impossible for the lack of a shared secret key.

A compromise can be valuable in such situations. A *hybrid cryptosystem* uses a public key cryptosystem once at the beginning of the communication to share a short piece of secret information that is then used as the key to encipher and decipher messages by means of a "conventional" secret key cryptosystem. If the message is sufficiently long, it is preferable to use the public key cryptosystem several times during the transmission, so as to change secret keys often. Without much slowing down the protocol, this greatly increases the safety of the hybrid system for two reasons: it is easier to break a secret key system if much ciphertext is available and, even if the cryptanalyst succeeds in figuring out one of the secret keys, he can only decipher the corresponding part of the message.

5 Applications

5.1 Password Protection

Access control to computer systems is generally based on a small piece of confidential information, a so called *password* which is shared between the user and the computer system. The latter can request this information whenever the user's identity is questionable. Clearly, the value of such an access control depends crucially on the ability to keep passwords secret. The mere fact that passwords are available in some table to the computer is very dangerous. No matter how well they are hidden or protected, a clever enemy can always find them wherever they are. Moreover, in most computer environments, there is at least one person, the system administrator, who has legal access to the password directory.

This threat can easily be overcome in practice. The main idea is that the computer does not really need to know the actual passwords; it merely needs to be capable of validating a given password. This can be implemented through the use of one-way functions. It is enough for the computer to store the images of the users' passwords under some one-way function. Because it is one-way, there is no point in keeping the function itself secret. Whenever a user wishes to prove his identity, he transmits his actual password on which the computer immediately applies the one-way function. The result can be validated against the computer's table.

This scheme can be rather weak if the users are likely to use easy-to-guess passwords. The following attack can often be mounted successfully by someone who has access to the table of images of the passwords and to the one-way function. Choose the most common passwords, compute the one-way function on them, and efficiently compare the results with the table. This attack does not allow to selectively find the password of a user, but it is very effective at finding several of them.

Although it does nothing additional towards protecting individual passwords, *salting* is a technique that nearly eliminates the above threat. The system chooses a random bit string for each user. The password table collects with each username the corresponding bit string (in clear) together with the value of the one-way function applied on the concatenation of the user's actual password with this random string. As a result, two users who may have chosen the same password will have distinct entries in the table.

5.2 Digital Signature Schemes

Authentication schemes based on secret key algorithms are perfectly suitable between two mutually trusting entities. They cannot be used to settle a possible dispute between them, in other words they do not allow to convince a third party of the authenticity of either message or sender. The advent of trapdoor one-way functions has allowed the introduction of the stronger notion of *digital signature*. If A sends a digitally signed message to B, then B will not only be convinced that the message was signed by A but he will also be able to prove to a third party that A actually signed that message. Note that a digital signature proofs as well the authenticity of the message as the authenticity of the sender. A public cryptosystem as introduced in Sect. 3.3 offers digital signature capacity if

$$E_{k_e}[D_{k_d}(M)] = M$$

for every message M, that means if the keys "commute".

Let A be a user with secret key s_A and corresponding public key p_A used with a public key cryptosystem with enciphering algorithm E and deciphering

algorithm D. Then A's signature on a message M is calculated as $S = D_{s_A}(M)$ while anybody knowing A's public key p_A can verify the signature by checking that $E_{p_A}(S) = E_{p_A}(D_{s_A}(M))$ equals M. In this way a third party can be convinced that no one but A could have signed the message M. In other words, the private deciphering algorithm of a public key system can be considered as a signature algorithm, while the public enciphering algorithm can be considered as the corresponding signature verification algorithm.
The digital signature is shown in Fig. 3.

$$A\,(s_A, p_A) \qquad\qquad\qquad B$$

$$S = D_{s_A}(M) \longrightarrow$$

$$E_{p_A}(S) =$$
$$E_{p_A}(D_{s_A}(M)) = M$$

Fig. 3. Digital signature scheme

Digital signatures can be used in conjunction with public key encryption if secrecy of the message is also desirable. Assume that users A and B have key pairs (s_A, p_A) and (s_B, p_B) respectively. If A wants to send the signed message M privately to B, she uses the signature algorithm D with her secret key s_A and encrypts the result applying the algorithm E with B's public key p_B. The result $C = E_{p_B}(D_{s_A}(M))$ is transmitted to B. The receiver B will first decrypt the message received by applying the decryption algorithm D with his secret key s_B and then verify A's signature by applying the verification algorithm E with A's public key p_A. This scheme is illustrated in Fig. 4.

$$A\,(s_A, p_A) \qquad\qquad\qquad B\,(s_B, p_B)$$

$$C = E_{p_B}(D_{s_A}(M)) \longrightarrow$$

$$E_{p_A}[D_{s_B}(C)] = E_{p_A}[D_{s_B}[E_{p_B}(D_{s_A}(M))]]$$
$$= E_{p_A}[D_{s_A}(M)] = M$$

Fig. 4. Digital signature scheme with encryption

The natural alternative would be for A to first encrypt the message M and then sign it afterwards. This is to be avoided because it would allow an eavesdropper to verify A's signature and to sign the encrypted message himself. The receiver B would not know that the eavesdropper was not the original sender of the message.

Since public key cryptosystems are slow, the digital signature on a large message is not calculated on the message itself but on a hash value obtained from the original message. This hash value is the result of the application of a so called *hashing function* on the message. A hashing function h is a publicly known one-way function with the property that it is *collision free* which means that it is computationally infeasible to find different messages M and M' such that the hash values $h(M)$ and $h(M')$ are equal. Note that when a hash function is applied before signing the message M, it is necessary to send the message M together with the signature since it will be needed in the verification process to recalculate the hash value as shown in Fig. 5.

$$A\ (s_A, p_A) \qquad\qquad\qquad\qquad B$$

$$S = D_{s_A}[h(M)],\ M \longrightarrow$$

calculates $h(M)$ and
$$E_{p_A}(S) =$$
$$E_{p_A}(D_{s_A}[h(M)]) = h'(M)$$

checks that $h(M) = h'(M)$

Fig. 5. Digital signature scheme with hashing function

Digital Signature Based on RSA. The digital signature scheme based on the RSA cryptosystem (see Sect. 3.4) is the most famous one. However, if RSA is used both for privacy and signatures, it is preferably that each user keeps two distinct pairs of keys for the two distinct purposes. This separation serves two purposes. Generally to speed up the verification of a signature only small public exponents are used with digital signatures. However encryption with this small exponents might be rather insecure [6]. On the other hand we have seen in Sect. 3.3 that public key cryptosystems are weak under chosen ciphertext attacks. Such attacks can be harder to mount if signing is different from deciphering.

ElGamal Digital Signature. In 1985 ElGamal [12] introduced a new digital signature scheme based on the discrete logarithm problem.

Let p be a large prime number and a in $[2, p-1]$. Choose a secret key x in the same range and calculate the public key $y \equiv a^x \bmod p$. Let M be a message in $[0, p]$.

The signature process on M is as follows:

– choose a random k in $[0, p-1]$ with $\gcd(k, p-1) = 1$

- compute $r \equiv a^k \bmod p$
- the signature for M is the pair (r, s) with s the solution of

$$a^M \equiv a^{x \cdot r} \cdot a^{k \cdot s} \bmod p$$

or

$$M \equiv (x \cdot r + k \cdot s) \bmod p - 1$$

or

$$s \equiv (M - x \cdot r) \cdot k^{-1} \bmod p - 1,$$

which has a solution since $\gcd(k, p - 1) = 1$.

For the verification one has to check that for given M, r, s, there holds

$$a^M \equiv y^r \cdot r^s \bmod p,$$

since this equals $a^M \equiv a^{x \cdot r} \cdot a^{k \cdot s} \bmod p$.

Notice that the parameter k should not be used more than once and that the parameter p should have at least one large prime factor. If $p - 1$ has only small prime factors, then computing the discrete logarithm is easy. A discussion on the possible attacks to this scheme is beyond the scope of this paper. Most of them are easily shown to be equivalent to computing discrete logarithms over $GF(p)$. We refer the interested reader to [12] for further information on the security of the scheme.

References

1. L. Adleman, *On breaking the iterated Merkle-Hellman public key cryptosystem*, Advances in Cryptology, Proceedings of Crypto '82, Plenum Press, 1983, 259–265.
2. G. B. Agnew, R. C. Mullin and S. A. Vanstone, *Fast Exponentiation in $GF(2^n)$*, Advances in Cryptology, Proceedings of Eurocrypt '88, Lect. Notes Comp. Science 330, 1988, Springer Verlag, 251–254.
3. D. W. Ash, I. F. Blake and S. A. Vanstone, *Low complexity normal bases*, Discr. Applied Math. 25 (1989), 191–210.
4. I. F. Blake, R. Fuji-Hara, R. C. Mullin and S. A. Vanstone, *Computing logarithms in finite fields of characteristic two*, Siam J. Alg. Disc. Math., Vol. 5 nr. 2 (1984), 276–285.
5. I. A. Blake, P. C. van Oorschot and S. A. Vanstone, *Complexity Issues for Public Key Cryptography*, in Performance Limits in Communication Theory and Practice, J.K. Skwirzynski, Ed., Kluwer Academic Publishers, 1988, 75–97.
6. M. Blum, *A potential danger with low-exponent modular encryption schemes: avoid encrypting exactly the same message to several people*, preprint (1983).
7. G. Brassard, *Modern Cryptology. A Tutorial*, Lect. Notes Comp. Science 325, 1988, Springer Verlag.
8. D. Coppersmith, A. M. Odlyzko and R. Schroeppel, *Discrete Logarithms in $GF(p)$*, Algorithmica 1 (1986), 1–15.
9. D. E. Denning, *Cryptography and Data Security*, Addison–Wesley Publ., 1983.

10. Y. Desmedt, J. Vandewalle and R. Govaerts, *Critical analysis of the security of the knapsack public key algorithms*, IEEE Trans. Inform. Theory, Vol. IT–30 (1984), 601–611.

11. W. Diffie and M. E. Hellman, *New Directions in Cryptography*, IEEE Trans. Inform. Theory, Vol. IT-22 (6) (1976), 644–654.

12. T. ElGamal, *A Public Key Cryptosystem and a Signature Scheme Based on Discrete Logarithms*, IEEE Trans. Inf. Theory, Vol. 31–4 (1985), 469–472.

13. N. Koblitz, *A Course in Number Theory and Cryptography*, Graduate Texts in Math. 114, Springer-Verlag, New York, Berlin, Heidelberg, 1987.

14. D. E. Knuth, *The Art of Computer Programming*, Addison-Wesley Reading MA., Vol. 3, 1973.

15. A. K. Lenstra, H. W. Lenstra, Jr., M. S. Manasse and J. M. Pollard, *The number field sieve*, preprint.

16. R. C. Merkle and M. E. Hellman, *Hiding information and signatures in trapdoor knapsacks*, IEEE Trans. Inf. Theory, Vol. IT-24 (1978), 525–530.

17. R. C. Merkle, *Protocols for Public Key Cryptosystems*, Proceedings 1980 Symp. on Security and Privacy IEEE Comp. Soc., (1980), 122–133.

18. G. L. Miller, *Riemann's hypothesis and tests for primality*, Proceedings of the Seventh Annual ACM Symposium on the Theory of Computing, 234–239.

19. R. C. Mullin, I. M. Onyszchuk and S. A. Vanstone, *Optimal normal bases in $GF(p^n)$*, Discrete Applied Math. 22 (1988/89), 149–161.

20. A. Odlyzko, *Discrete Logarithms in Finite Fields and their Cryptographic Significance*, Advances in Cryptology, Advances in Cryptology, Proceedings of Eurocrypt '84, Lect. Notes Comp. Science 209, 1985, Springer Verlag, 224–314.

21. J.-J. Quisquater and C. Couvreur, *Fast decipherment algorithm for RSA public key cryptosystem*, Electronic Letters, Vol. 18 (1982), 905–907.

22. M. O. Rabin, *Probabilistic algorithms for testing primality*, J. of Number Theory, Vol. 12 (1980), 128–138.

23. R. Rivest, A. Shamir and L. Adleman, *A method for obtaining digital signatures and public key cryptosystems*, Comm. ACM, Vol. 21 (1978), 120–128.

24. R. Rivest, *RSA Chips (Past / Present / Future)*, Advances in Cryptology, Proceedings of Eurocrypt '84, Lect. Notes Comp. Science 209, 1985, Springer Verlag, 159–165.

25. A. Shamir, *A polynomial time algorithm for braking the basic Merkle-Hellman cryptosystem*, Proceedings of the 23rd IEEE Symposium on Foundations of Computer Science, 1982, 145–152.

26. P. C. van Oorschot, *A comparison of practical public key cryptography based on integer factorization and discrete logarithms*, Advances in Cryptology, Proceedings of Crypto '90, Lect. Notes Comp. Science 537, 1991, Springer Verlag, 576–581.

Better login protocols for computer networks

Dominique de Waleffe and Jean-Jacques Quisquater[1]

[1] UCL, Dept. of Electrical Engineering,
Place du Levant 3, B-1348 Louvain-la-Neuve.

quisquater@dice.ucl.ac.be

Abstract

Authenticating computer users is a fairly old problem. Password based
solutions were acceptable until the growth of computer networks based on
insecure communication. Today many systems still use fixed passwords as
a means of authentication. We show in this paper how an old scheme by
Lamport can be used to provide more security. Relying on that scheme and
zero-knowledge techniques, we show extensions providing much more general
access control mechanisms.

Those extensions can be exploited in several ways: to authenticate users
in computer networks, to provide users with access tickets or provide servers
with proofs of usage.

We also show how, in a single transaction, a user can prove this authenticity
as well as prove his possession of a ticket.

Finally, we explain how smart cards make those protocols very practical.

1 Introduction

Computer systems managers have always wanted to control access to the machines, both for security and management reasons. In the old days of computers, machines where situated at the same site as users; often there was only one single machine accessible to people, information sharing was mostly controlled by systems managers. Thus a single login control was sufficient for most purposes.

Nowadays, computers are connected via networks, people routinely use services provided by remote computers; however techniques to authenticate users across those networks are still often the old ones. To achieve security and management objectives, computer systems managers have always imposed access restrictions to the systems. Sometimes physical restrictions were enough (only a few selected individuals had the computer room key); later, when computer users were growing in number but were still located on the same site, a simple login protocol was sufficient to authenticate users.

Today, even though computers are connected by insecure, large, networks and users routinely access services provided by remote computers, authentication is still largely based on techniques offering very little confidence.

We define *login problem* as that of identifying with confidence users accessing a computer or a computer based service.

In this paper, we examine available techniques that, to a limited extent, solve the login problem, then we introduce protocols which, without being too complicated or costly, offer more confidence in today's complex networked environments. They could be used in the same context as Kerberos [6] and similar systems, but they also solve some problems associated with these systems.

Let us first define a few terms used in this paper. Computer systems are means to provide users with services. Those services are represented by programs which are the working entities; we call those programs *servers*. In order to grant access to users, the servers have to verify some properties, hence we also call them *verifiers*. On the other side, users have to demonstrate their identity, thus we call them *provers*. There are three other parties involved in the treatment of security, they are a *trusted third party*, an *authority* and *opponents* or malicious entities.

The authority is a party[1] that manages and controls the security system, the authority also distributes control information to participants; in some cases, the authority initializes the system then disappears.

Depending on the requirements of the policy and/or security system, the trusted third party serves as an arbiter, to resolve conflicts, or as a notary, to record critical information independently of the other parties.

The opponents can be wire-tappers listening to or modifying the traffic on communication lines or they can be active insider entities like malicious programs (Trojan horses).

Section 2 briefly examines families of protocols in use, identifies their properties and derives a number of other desirable properties.

A particular implementation of an old scheme, introduced by Lamport, is presented in section 3. In section 4, zero-knowledge techniques are combined with this protocol in order to provide higher security, flexibility and additional properties such as undeniable proofs of usage. In section 5, identity-based systems are briefly presented. Those can be combined with our ticket mechanism to provide a common framework for both user and ticket authentication (section 6). Extensions for multiple networks and continuous authentication are introduced in section 7 and some implementation considerations are also given in section 8.

---

[1] The authority can be a person or a secured computer.

53

2 Current protocols

2.1 Fixed Passwords

A widely used solution is based on fixed passwords. The computer maintains a list of known users with their password (in clear or encrypted). In order to be granted access, the user must type his password at the terminal which transmits it to the computer (figure 1).

```
1. User: sends identity,

2. Server: requests password,

3. User: sends password,

4. Server: verifies that the given password
   corresponds to the one in the table.
```

Figure 1: Fixed password protocols

This family of solutions has several problems:

- Passwords are transmitted in clear over the communication lines and are therefore subject to copying by eavesdroppers or malicious communication programs.

- Password files contain very sensible information and may be subject to a number of attacks. This forces managers to implement high security policies for these particular files.

- Correct authentication of users across networks is quite difficult as password file information has to be disseminated to remote machines.

- To ensure that passwords cannot be easily guessed by intruders the use of uncommon and therefore hard to remember passwords is necessary.

2.2 Shared key

A number of problems can be solved by a secret sharing scheme where the authentication protocol is different at every occurrence.

Both parties somehow agree on a common secret K. To authenticate the user, the server sends a non predictable value r to the user who computes $p = F_K(r)$ where F is a public enciphering function (DES for example) parameterized by K. The user sends the ciphertext p to the server which can also compute $F_K(r)$ and compare both values. In this scheme, no secret information is transmitted, the computation time is constant for each login. However, as all parties know the secret, impersonation of a party by another is possible.

K is a shared secret key for enciphering function F.

1. User: *sends identity,*

2. Server: *sends control value r and finds K for the user,*

3. User: *sends password $p = F_K(r)$,*

4. Server: *verifies that received $p = F_K(r)$ by also doing the computation.*

Figure 2: Shared key

This protocol has been implemented in a number of products which can be easily purchased.

Other schemes are based on having several secret signature functions owned by the authority some of which are shared with verifiers [13].

Those schemes require that a common key be agreed on to start with. One can use the Diffie-Hellman scheme or adopt a protocol based on master keys shared by the authority and the verifiers [13].

2.3 Lamport's scheme

A better approach, also based on using a different password at each access, eliminates the necessity for a shared secret. Only the prover has a secret, and thus cannot be impersonated.

In [5] Lamport describes a very elegant authentication scheme based on a one-way function. Let F be a *one-way function*, that is, for a given x, the value $y = F(x)$ is easy to compute; while given y, it is computationally difficult[2] to find a value x such that $y = F(x)$.

We will note x_i the value computed from the following rules:

$$x_0 = \mu_0 \text{ is an initial value,}$$

$$x_i = F(x_{i-1}) \text{ for } i = 1, 2, 3, \ldots$$

The scheme goes as follow (figure 3). The one-way function is public; the user chooses an initial $x_0 = \mu_0$ as the starting value and keeps it secret; then he gives x_{1000} to the verifying entity[3], there is no need to keep that value secret.

---

[2] More exactly, a one-way function is impossible to invert, but the computational difficulty assumption is enough for practical purposes. A trap-door one-way function is somewhat similar except for the existence of some secret which allows the inversion. An example is: the knowledge of the factorisation of n into p and q allows one to extract roots modulo n.

[3] The value 1000 is an example, any other suitable value can be chosen.

To be recognized, the user will be asked to deliver x_{999} which can be computed easily from x_0, kept secret. Due to the one-way property of F, nobody except the authentic user can show x_{999}.

Now, as x_{999} has been disclosed, it cannot be used anymore. That is no problem, the verifier remembers x_{999}, which he just received, in place of x_{1000} and will expect x_{998} as the next password. This can continue until x_0 is finally used as password. Therefore a new password is expected at each login. The value retained by the verifier can be seen as the *current identity* of the user, as it can be presented by him and only he can also show the corresponding password.

A necessary condition for the security of the scheme is that all x_i's must be different, moreover, if a too small number of iterations of the function can give as a result some previously obtained value, the system may not be secure. The number of possible values must be large enough to prevent attacks based on these easy forward computations.

---

1. **User**: *sends current identity* x_k,

2. **Server**: *verifies that* x_k *is known then request password* $x_{k-1} = F^{-1}(x_k)$,

3. **User**: *sends password* x_{k-1},

4. **Server**: *verifies that* $x_k = F(x_{k-1})$; *then remembers* x_{k-1} *as identity for next access.*

Figure 3: Lamport's protocol

This scheme provides a limited number of login possibilities, however it does not rely on an authority and requires only limited secret storage. There is however a slight performance inconvenience: the sequence must be recomputed at each login; this can be circumvented by storing a number of intermediate values in the chain of x_i's.

The main problem with the scenario is that the value x_{1000} must be certified to have come from a valid user. For instance, this could be provided by requiring physical presence of the user to initialize the scheme. This is not different from any other scheme where initialization requires the user to be authenticated.

Implementations have been neither very practical nor very secure (based on the file system security) until recently with the emergence of smart cards [10] able to compute and memorize the future password. Smart cards can be used to store the user's secret μ_0 and do the computations of x_k at login time. Additionally, when the login transaction is complete, the card can precompute and store the password for the next login.

2.4 Tickets

The idea of access tickets for networked services has also found its place in a practical implementation in the authentication service Kerberos [6]. A *ticket* is some information, computed by a trusted third party (Ticket Granting Server or TGS in Kerberos terms), which the user shows to the server he wants to access. Due to ticket structure, the verifying entity can determine the validity and authenticity of the ticket, then grant access. Kerberos tickets have a validity limited by time stamps.

In Kerberos, however the initial login is still password based and as such is vulnerable to password attacks.

2.5 Desirable properties

From those protocols and their properties one can infer a number of qualities which login protocols for networks should have:

- No secret should be transmitted over communication lines, to avoid stealing.

- No secret should be stored on the verifier side, to avoid having to secure it.

- If a secret must be known by the verifier, it must be different from the user's secret to avoid impersonation.

- The scheme should allow a server to have a proof of usage by users.

- A mutual authentication protocol should be easily implemented within the same framework.

- Protocol should not take too long and should not impose too much of a burden on users (like 100 characters passwords, ...).

3 An instance of Lamport's scheme

When the integer n satisfies adequate conditions, extracting roots in Z_n, the ring of integers modulo n, is a difficult problem while computing powers, the inverse function, is very easy.

Let $v \geq 2$, $n = p \cdot q$ with p and q two large safe primes[4] approximately the same length (≈ 250 bits), the function

$$F(x) : x^v \bmod n$$

must be considered as a one-way function by anyone not knowing p and q.

The function F can be used to build an instance of Lamport's protocol. An authority is introduced to have some control over the system: the authority computes the parameters and is responsible for the maintenance of the relationships between users' real identities and their identities as seen by the different servers. Further, the authority knows the trap-door (the factors of n) and is therefore able to trace users trying to cheat. This function cannot be delegated to each server in a large system.

The scheme can be summarized as follows, assuming one verifier:

---

[4] A prime p is said to be safe if $p = 2 \cdot p' + 1$ where p' is also prime. A product of safe primes is in general more difficult to factor.

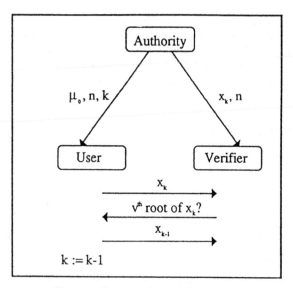

Figure 4: Lamport's modified scheme

- The authority computes a suitable n and generates a good μ_0 for each user. The exponent v is assumed to be known by the parties.

- The authority transmits μ_0, n and k, the number of available tickets to the user. The value μ_0 must be kept confidential during this transaction.

- The authority provides the server with n, the modulus and x_k, the identity of a possible user.

- The user can access the server by showing x_k as identity, the verifier checks that indeed, x_k represents one valid identity. Then the verifier asks as password the vth root of x_k. This value is x_{k-1} and can only be computed by the user from μ_0. The verifier can easily check the password validity by computing $x_{k-1}^v \bmod n$ which must be x_k.

- The password x_{k-1} (if correct) becomes the identity for the next access.

The protocol is illustrated in figure 4.

As this function F will be used repeatedly, it is not impossible that a cycle exists. The number of possible values obtained by successive evaluations of x_i's must be assessed (see appendix).

In this scheme, the authority provides a user with k access tickets to a service and gives the service means for checking and consuming the tickets. Each ticket represents the user's current identity for the service but it is not an absolute identity, which is maintained by the authority in relations like (user id, μ_0, service id). Additionally, the authority is the only one able to extract v^{th} roots, therefore only the authority is capable of determining the real identity of a cheater using some authentic x_k in a fraudulent transaction.

As presented here, this scheme has drawbacks: tickets can be used only once, they must be disclosed before using the service and in case of server failures, some old tickets may be replayed. These drawbacks may lead to many transactions with the authority to get new access tickets; they may allow a cheater to use x_{k-1} in an illegal transaction if, for some reason, the service has not been provided and the verifier (which has crashed and restarted, for instance) has only remembered x_k as current identity; finally, having shown your ticket does not imply you have received or been satisfied with the wanted service.

4 A better protocol

Informally, a zero-knowledge protocol is a dialogue during which two entities, a prover and a verifier, exchange information and at the end of which the verifier is convinced that the prover knows a secret without that secret being revealed whatsoever. Further, an observer cannot be convinced as the verifier is able to compute himself (in reverse order) all the elements of the protocol.

Combining such protocols with the previous scheme, i.e. proving knowledge of the current password without revealing it, circumvents two of its drawbacks:

- The current password is not revealed during the transaction, therefore it can be used several times.

- The current password may be revealed only after a successful usage of the server.

The new protocol remains very simple (figure 5). Instead of sending the current ticket x_{k-1}, the prover (user) convinces the verifier that he knows x_{k-1}. Being convinced, the verifier can grant access, then finally the verifier may ask the user to disclose the ticket or may let the user keep the same ticket for the next access. Of course, upon reception of x_{k-1} the verifier will check that $x_k = x_{k-1}^v \bmod n$ to ensure the authenticity of the identity used in the next access, that is x_{k-1}.

There are advantages to such schemes:

- Without resorting to time-stamps, the verifier is able to fix an appropriate lifetime to individual tickets.

- If a ticket is consumed after each usage of the service, the ticket can be used as a proof that the service was granted to the user. The user cannot deny granting of the service as he is the only one able to provide such a valid ticket.

- No secret information is revealed during transactions; when x_{k-1} is transmitted, it may be public.

We will now detail the complete protocol for two well known zero-knowledge protocols: that of Fiat and Shamir [2] and the more recent one published by Guillou and Quisquater [3].

> 1. **User:** *sends x_k as identity,*
>
> 2. **Server:** *asks a proof of knowledge of x_{k-1},*
>
> 3. **User:** *sends a public convincing argument,*
>
> 4. **Server:** *grants access,*
>
> 5. **Server:** *later, asks ticket x_{k-1},*
>
> 6. **User:** *sends x_{k-1}.*

Figure 5: Lamport's scheme with zero-knowledge protocol

4.1   An example of Fiat-Shamir protocol

The user wants to prove to the verifier that he knows the vth root of x_k. Both know the modulus n. The protocol goes as follows (figure 6):

1. The user sends his current identity x_k and his commitment $T = r^v \bmod n$ where r is a user generated random number.

2. The verifier sends a binary question d.

3. The user sends back either $t = r$ if $d = 0$, or $t = r \cdot x_{k-1} \bmod n$ if $d = 1$.

4. The verifier checks the validity of the reply by computing $t^v \bmod n$ which must be either $T = r^v \bmod n$ or $r^v \cdot x_{k-1}^v \bmod n = r^v \cdot x_k \bmod n$.

5. Repeat steps 1 to 4 a number of times to reach the desired confidence level.

At each repetition of the protocol, the verifier doubles his confidence in the prover, thus the protocol must be repeated enough times to reach a sufficient level of security. There are possible optimizations (see, for example, appendix of [12]).

Why is this zero-knowledge? Informally, T is a random number, except if one knows the factors of n, one cannot find r, therefore when the prover transmits $t = r \cdot x_{k-1}$, that value appears to be random and nobody can retrieve x_{k-1}. Thus all transmitted information appears to be random and does not reveal the secret.

The verifier is convinced after a number of rounds as, for each round, the prover has 50% chances of guessing the question d correctly and thus preparing a correct $(T = r^v \cdot x_k^d, t = r)$ pair.

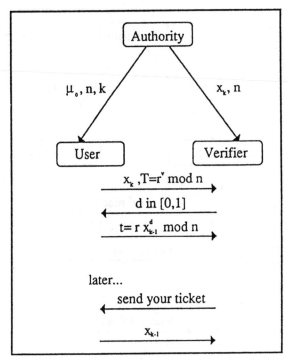

Figure 6: Ticket authentication with F-S protocol

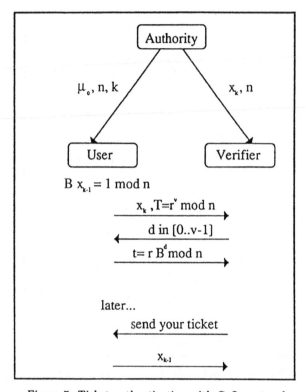

Figure 7: Ticket authentication with G-Q protocol

4.2   Guillou-Quisquater protocol

One drawback of using the F-S protocol is the repetition of the protocol necessary to reach a given security level. Using the G-Q protocol implies

- One more computation for the user, however this need be done only once per ticket x_i.

- A slightly more complex verification computation.

- Only one round is necessary to reach a given security level.

In this case, the protocol goes as follows (figure 7), where v is chosen according to the desired confidence level:

1. The user computes B such that

$$B \cdot x_{k-1} \bmod n = 1, \tag{1}$$

the extended Euclidean algorithm can be used to compute B efficiently. This further implies that $B^v \cdot x_k \bmod n = 1$.

2. The user sends his current identity x_k and his commitment $T = r^v \bmod n$ where r is a user generated random number.

3. The verifier sends a random question d such that $0 \le d \le v - 1$.

4. The user sends back $t = r \cdot B^d \bmod n$.

5. The verifier checks the validity by computing

$$
\begin{aligned}
x_k^d \cdot t^v \bmod n &= x_k^d \cdot r^v \cdot B^{d^v} \bmod n \\
&= x_{k-1}{}^{d^v} \cdot B^{d^v} \cdot r^v \bmod n \\
&= r^v \bmod n, \text{ due to relation (1)} \\
&= T.
\end{aligned}
$$

In this protocol, the user has one chance in v to guess the question d and precompute a valid (T, t) pair. Thus with v large enough, a given security level can be reached in a single transaction.

5 Identity-based systems

In password based schemes, the users' identities are fixed at all times, thus provide a fixed reference. In Lamport's scheme, the representation of a user's identity changes every time a ticket is consumed. Further, there is no absolute need for an authority in such schemes, as servers can generate μ_0 and compute the initial identity for each of its users. In those cases however, each server should then be able to authenticate the receivers of the initial tickets.

We now briefly outline a system in which each user has a different but fixed identity I, which is validated once by the authority at the beginning, and in which, servers can authenticate the users without knowledge of any secret.

Identity-based systems [11] are a consequence of zero-knowledge protocols. Within such a system, the authority knows a modulus $n = p \cdot q$ whose factors are kept secret. Each user has his identity I. When the system is initialized, every registering user's identity is signed by the authority which computes $S(J)$ where $J = Red(I)$ is a public redundancy function and $S(J) = J^{1/v} \bmod n$. This function S is only computable if the factors of n are known, i.e. only by the authority.

The signatures are kept in in users' smart cards, and thus remain secret. All the provers and verifiers must know v and n, then the system can be closed by removing the authority.

In order to be authenticated by a verifier, a user shows his identity and proves by a zero-knowledge protocol that he knows $S(J)$ without revealing it.

Such a system requires one secret for each prover, only the authority knows the factorization of n, no secret is needed on the verifier's side. Further there is only one modulus for the whole system.

6 Combining two mechanisms

A family of practical *cooperation protocols* was introduced by Guillou and Quisquater in their CRYPTO '88 paper [4]. They allow two proofs to be combined into a single one, not much more complex. For instance, they show how to prove the simultaneous presence of two users having identities I_1 and I_2 given a single modulus and exponent. Those protocols can be used to combine the identity based and ticket based mechanism so that a single proof demonstrates both possession of a valid ticket and identity.

The setup is the following:

- The authority owns a modulus $n = p \cdot q$, where n is public while its factors are secret.

- Each user has an identity I and a signature of this identity $S(J)$.

- Each user is provided an initial pair (μ_0, k) pair for the service, that is k access tickets.

- The service (verifier) is provided with x_k and modulus n.

- The exponent v is common for user and ticket authentication.

Let us detail this combined protocol in the context of G-Q proofs.

The user computes once a B_1 such that

$$B_1^v \cdot J \bmod n = 1, \text{ with } J = Red(I).$$

To prove his possession of $S(J)$ and a ticket x_{k-1}:

1. The user computes B_2 such that
   $B_2 \cdot x_{k-1} \bmod n = 1$.

   The user picks a random r_1 in Z_n^* (i.e. Z_n without 0) and computes $T_1 = r_1^v \bmod n$.

   The user picks a random r_2 in Z_n^* and computes $T_2 = r_2^v \bmod n$.

   The user sends I, x_k and his commitment T computed from:

   $$\begin{aligned} T &= T_1 \cdot T_2 \bmod n \\ &= (r_1 \cdot r_2)^v \bmod n \\ &= r^v \bmod n. \end{aligned}$$

2. The verifier sends a random question d between 0 and $v - 1$.

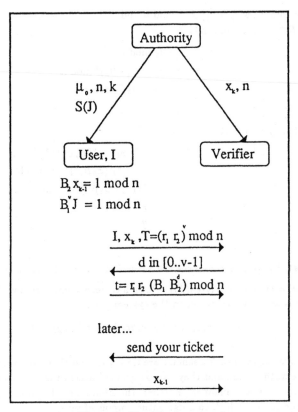

Figure 8: Combined protocol

3. The user computes $t_1 = r_1 \cdot B_1{}^d \bmod n$ and
$t_2 = r_2 \cdot B_2{}^d \bmod n$ and sends

$$
\begin{aligned}
t &= t_1 \cdot t_2 \bmod n \\
&= r_1 \cdot r_2 \cdot (B_1 \cdot B_2)^d \bmod n \\
&= r \cdot (B_1 \cdot B_2)^d \bmod n.
\end{aligned}
$$

4. The verifier checks that

$$
\begin{aligned}
J^d \cdot x_k{}^d \cdot t^v \bmod n &= J^d \cdot x_k{}^d \cdot r^v (B_1 B_2)^{v \cdot d} \bmod n \\
&= (J B_1{}^v)^d (x_k B_2{}^v)^d \cdot r^v \bmod n \\
&= (x_{k-1} \cdot B_2)^{(v \cdot d)} \cdot r^v \bmod n \\
&= r^v \bmod n \\
&= T.
\end{aligned}
$$

At little cost, the prover convinces the verifier of two unrelated facts:

- He knows a signature $S(Red(I))$ which has been computed by the authority, therefore is an authorized user in the system.

- He owns an access ticket x_{k-1} for the server he is talking to, thus can be granted access.

The main advantages of such a scheme are that with a single modulus and exponent, all users can be authentified and they can be granted access tickets to any number of services in an independent manner. Further, the lifetime of the tickets can be easily controlled by the servers which can also gather proofs of usage.

A further point is: the prover does not disclose any part of its secret in doing so.

The protocol can be extended to include cases where there is an exponent v_1 used for the identity verification and v_2 used for the tickets. This scheme allows different security levels in a single proof. The following modifications need to be applied to the protocol:

1. Let $v = v_1 \cdot v_2$.

2. The user computes B such that $B \cdot x_{x-1} \cdot S(J) = 1 \bmod n$, this implies $B^{v_1 \cdot v_2} \cdot J^{v_2} \cdot x_k^{v_1} = 1 \bmod n$.

3. The user sends $T = r^v \bmod n$, where r is a random integer greater than 0.

4. The verifier sends a random question d between 0 and $v - 1$.

5. The user sends $t = r \cdot B^d \bmod n$.

6. The verifier checks that $J^{v_2 \cdot d} \cdot x_k^{v_1 \cdot d} \cdot t^v = T \bmod n$.

In addition, this extension can be used to prove ownership of two independent tickets which are needed to access a particular server.

7 Extensions

7.1 Multiple networks

The protocols we have mentioned can be used in quite large networks, however, a single network managed by a single authority with a unique modulus would appear like Big Brother, thus is not acceptable. Rather, while all organizations want to keep control of their computer networks and users, they are willing to share some resources with others without yielding control on security issues. Thus the question is: can we extend those schemes to several connected networks?

We are thus in the following situation (taking two networks as an example):

- in network 1, the modulus is $n_1 = p_1 \cdot q_1$ and the exponent is v_1,

- in network 2, the modulus is $n_2 = p_2 \cdot q_2$ and the exponent is v_2,

- user I_1 of network 1 has been granted access tickets x_k to a server on network 2.

Provided n_1 and n_2 are relatively prime[5], one can use the previous protocols but do the computations in $Z_{n_1 \cdot n_2}$ and use the Chinese Remainder theorem to fasten the computations.

This allows a user of one network to prove to a server on another network that:

- he owns a signature of his own authority,

- he owns access tickets to that server on the other network.

7.2 Continuous authentication

Lamport's scheme can be used in an extension of the login protocol to authenticate all commands sent by the user to the server.

Each command is sent enciphered using the current ticket as key, then the ticket is sent. The server checks the validity of the ticket and may now use it to decipher the command, which is thus authentified since only the user can have constructed the key.

Such a protocol ensures an authentication during the whole session, which can be stopped at any time and restarted at the same point.

In the same vein but without consuming a ticket for each command, commands sent by users and replies sent by servers can each be accompanied by one part of a zero-knowledge protocol proving knowledge of the ticket. This ensures that the whole session is with the same user.

7.3 Mutual authentication

Both parties can be mutually convinced by adding one more message to the protocol. This property may be desirable. The thoughtful reader will easily make the extension.

---

[5] This condition is necessary anyway. Otherwise computing $gcd(n_1, n_2)$ would give the factorization for both and thus break the system.

> 1. **User:** *sends* $C = Encipher(cmd, x_{k-1})$,
>
> 2. **User:** *sends* x_{k-1},
>
> 3. **Server:** *checks* $F(x_{k-1}) = x_k$,
>
> 4. **Server:** *executes* $Decipher(C, x_{k-1})$.

Figure 9: Continuous Lamport authentication

> 1. **User:** *sends* (cmd, T),
>
> 2. **Server:** *sends* $(reply, d)$,
>
> 3. **User:** *sends* (cmd, t).

Figure 10: Continuous ticket authentication

8 Implementation aspects

The protocols described herein can be implemented in smart cards. The original protocol of Lamport could for instance use DES as a one-way function and therefore use current generation smart cards.

The newer protocols combining exponentiation and zero-knowledge require more powerful devices like the RSA smart card [1]. This particular card can do an exponentiation of 512 bits operands (exponent also) in 1.5 second. However an exponent of 60 bits provides a sufficiently high level of security; this means that only 0.2 second should be needed for one exponentiation.

The complete protocol including communication at 9600 bps with the card should take around 0.6 second (without network communication time, which can be either small or large). This seems like an acceptable delay. Further, the user would only have to type his PIN code when the card is inserted in the reader, everything should work transparently after.

As regards storage requirements, the identity, signature and modulus must be kept along with the exponent (1596 bits) plus one μ_0, x_{k-1} and k ($\approx$ 1050 bits) for each service.

The computations of μ_0 and the initial x_k are quite time consuming, however these could be done by a special purpose machine, which could compute very long sequences for good μ_0's and yet distribute only part of the sequence at a given time. One must note that the two parts of the sequence depending on one μ_0 may not be in use at the same time in the system.

9 Conclusion

We have presented the problems associated with a number of login protocols currently in use. We introduced a powerful, yet simple, protocol based on Lamport's scheme and zero-knowledge techniques. This protocol has the following characteristics:

- no secret is transmitted over communication lines,

- there is no shared secret,

- no unnecessary secret is kept on the verifier side,

- the scheme allows a server to have a proof of usage by users,

- mutual authentication can be easily added,

- the protocol can be readily used in computer networks,

- it is quite short and needs minimal interaction with the user,

- it can be implemented with existing technology.

We have also demonstrated extensions to several independent connected networks and to continuous authentication during sessions. Finally, we have explained why those protocol are practical.

References

[1] J.-J. Quisquater, D. de Waleffe and J.-P. Bournas, "*CORSAIR: A chip card with fast RSA capability*", Proceedings of Smart Card 2000, Amsterdam, 1989, to appear.

[2] A. Fiat, A. Shamir, "*How to prove yourself: practical solutions to identification and signature problems*", Proc. of CRYPTO '86, Lecture notes in Computer Science, Springer Verlag, Vol. 263, pp. 186–194.

[3] L. C. Guillou, J.-J. Quisquater, "*A practical zero-knowledge protocol fitted to security microprocessor minimizing both transmission and memory*", Proc. EURO-CRYPT '88, Lecture notes in Computer Science, Springer Verlag, Vol. 330, pp. 123–128.

[4] L. C. Guillou, J.-J. Quisquater, "*A 'paradoxical' identity-based signature scheme resulting from zero-knowledge*", Proc. CRYPTO '88, Lecture notes in Computer Science, Springer Verlag, Vol. 403, pp. 216–231.

[5] Leslie Lamport, "*Password Authentication With Insecure Communication*", C. ACM Volume 24, Number 11, pp. 770–772, Nov. 1981.

[6] J. Steiner, C. Neuman, J. Schiller, *"Kerberos: an authentication service for open network systems"*, Proc. of Winter Usenix '88, Dallas.

[7] J. Steiner, C. Neuman, *"Authentication of unknown entities on an insecure network of workstations"*, Proc. Usenix Security Workshop, Portland, Or, Aug. 1989.

[8] R. Needham, M. Schroeder, *"Using encryption for authentication in large networks of computers"*, C. ACM, Dec. 1978, pp. 993–999.

[9] M. Burrows, M. Abadi, R. Needham, *"A logic of authentication"*, Digital Equipment Corporation, Research Report, Feb. 1989.

[10] L. C. Guillou, M. Ugon, *"Smart Card: a highly reliable and portable security device"*, Proc. of CRYPTO '86, Lecture notes in Computer Science, Springer Verlag, Vol. 263, pp. 464–489.

[11] A. Shamir, *"Identity-based cryptosystems and signature schemes"*, Proc. of CRYPTO '84, Lecture notes in Computer Science, Springer Verlag, Vol. 196, pp. 47–53.

[12] L. C. Guillou, M. Davio, J.-J. Quisquater, *"Public-key techniques: randomness and redundancy"*, Cryptologia, Volume XIII, Number 2, pp. 167–189, Apr. 1989.

[13] J.-J. Quisquater, *"Secret distribution of keys for public-key systems"*, Proc. of CRYPTO '87, Lecture notes in Computer Science, Springer Verlag, Vol. 293, pp. 203–208.

[14] D. de Waleffe, J.-J. Quisquater, *"CORSAIR: A smart card for public-key cryptosystems"*, Proc. of CRYPTO '90, to appear.

[15] J.-J. Quisquater, L. C. Guillou & al., *"How to explain zero-knowledge to your children"*, Proc. of CRYPTO '89, Lecture notes in Computer Science, Springer Verlag, Vol. 435, pp. 628–631.

[16] K. H. Rosen, *Elementary number theory and its applications*, Addison-Wesley Publishing Co, 1984.

Appendix: Number of tickets for μ_0

As described in section 3, repeated evaluations of F starting from a publicly available value can loop back into values already obtained in the chain of the x_i's. This means that forward computations could reveal yet unused ticket authenticators.

The conditions enforced on n make this highly unlikely and costly as the following argument shows (the proofs for the necessary theorems can be found in [16]).

Definition 1 $ord_n a$ is the least integer x such that $a^x = 1 \bmod n$

Theorem 1 If a,n are relatively prime, $n > 0$ then $a^i = a^j \bmod n \Leftrightarrow i = j \bmod ord_n a$.

Theorem 2 *If a, n are relatively prime, $n > 0$ then $ord_n a \mid \phi(n)$.*

The initial μ_0 must be relatively prime to n which would otherwise be factored, the same must hold for μ_0^v.

A loop in the sequence of x_i's happens when, for some $i \neq j$, $x_i = x_j$ that is $\mu_0^{v^i} = \mu_0^{v^j} \bmod n$, thus (theorem 1) we need to show that $ord_n \mu_0^v$ is sufficiently large.

Theorem 2 says that $ord_n \mu_0^v$ is a divisor of $\phi(n)$. But n is a product of safe primes, thus $\phi(n) = 2 \cdot p' \cdot 2 \cdot q'$ where p' and q' are large primes. Hence $ord_n \mu_0^v$ is either 2, p' or q'.

To ensure that no short loops occur, one only needs to verify that $ord_n \mu_0^v \neq 2$ and this is a trivial operation.

Therefore with p around 256 bits, p' is around 255 bits, thus the number of tickets is of the order of 2^{255}. A forward computation would then require about 2^{255} large integer exponentiations.

A further result is that 1 cannot be reached in the sequence of x_i's if μ_0 is relatively prime to n.

The sequence of x_i's is the sequence $\mu_0^{0 \cdot v}, \mu_0^{1 \cdot v}, \mu_0^{2 \cdot v}, \mu_0^{3 \cdot v}, \ldots$

The smallest y such that $\mu_0^y = 1 \bmod n$ is $ord_n \mu_0$, that is, with p and q safe primes, one of 2, p' or q'. But v is not a divisor of p' or q' (and greater than 2 in practice), thus μ_0^y is not in the sequence.

Secret-Key Exchange with Authentication

Johan van Tilburg

PTT Research, P.O. Box 421,

2260 AK Leidschendam, the Netherlands

Abstract

This paper provides an outline for the second lecture on authentication protocols and deals with secret-key exchange protocols. The object is to encourage the interested reader to obtain and to study the original papers, and papers related to this subject.

1 Introduction

This second lecture on authentication protocols deals with the subject how to establish an authenticated secret key between two users on a public communication channel with public messages. In this situation a user must be able to verify, either directly or indirectly, that messages received have been originated by legitimate users and that they have not been altered or substituted for fraudulent ones. In this context, authentication can be viewed as a process between two legitimate users (sender and receiver) to ensure data integrity and to provide data origin authentication. It is important to note that secrecy and authentication are independent primitives. One can have perfect authenticity without secrecy and vice versa. We refer to [9] and [16] for a detailed treatment of this subject.

Public-key cryptography One of the benefits of public-key cryptosystems (PKS) in authentication protocols is that the public key does not have to be kept secret in order to have confidence in the authenticity of authenticated messages [16]. The security of PKS is based not on the secrecy of a shared secret key like in private-key cryptosystems (PrKS), but on the fact that it is generally believed to be computationally infeasible to derive the private key from a given public key. As consequence, in a PKS it is not necessary to exchange secret keys prior to communication. Because in a PKS the private key is not a shared secret, the users do not have to trust each other unconditionally (regarding the secrecy of the keys). For sending a message from user A to user B, A encrypts with B's public key. Only the legitimate user B can decrypt the message as only he owns the corresponding private key.

Because of its construction a PKS is often too slow or too expensive for most applications. This in contrast to a PrKS that has a low complexity and operates easily on speeds higher than 20 Mbit/s. For example, a provable secure PrKS (the one-time pad) is based on a simple exclusive-or operation, which adds (modulo 2) a message bit to a randomly generated secret key bit.

Shortly after the introduction of PKS, hybrid systems were proposed: The key exchange takes place with a PKS and the data encryption with a PrKS. However, for this kind of application it suffices to use a system for key exchange that constructs a secret. This type of system is called a public-key distribution system (PKDS) [4] and is used to establish secret keys between users.

Identification and authentication A proper identification scheme is a protocol that enables user A to prove his identity to user B without giving B the opportunity to misrepresent himself as A to someone else.

A PKS can be used to provide a proof of identity, for example, user A's knowledge or possession of his private key can be linked to his identity. In order to identify user A, one uses his public key to gain confidence that the user is in possession of the corresponding private key. A possible protocol runs as follows. Suppose user B (verifier) wants to verify the identity of user A (prover). The verifier sends a message (challenge) to the prover with the request to encipher. Prover A encrypts the message with his private key and returns the results to the verifier. Upon receipt, the verifier deciphers the message and compares it with the challenge sent. If the verification passes, the verifier is convinced (with high probability) that the prover is the legitimate user A who he claims to be, i.e., the prover possesses the private key associated with the claimed identity. Although the prover's public key is not secret, its integrity has to be guaranteed. In other words, when verifying identities of users with the help of PKS, one has to make sure that the used public key corresponds to the private key of the person who wants to prove his identity. Stated in terms of authentication channels, identity verification needs a public authentication channel to validate the public key of a private authentication channel belonging to the individual who wishes to prove his identity.

Advanced identification protocols make use of so called zero-knowledge identification schemes. In general, these schemes provide only authentication for an identity. Although outside the scope of this lecture, a zero-knowledge identification scheme that provides secret key exchange will be described in Section 4.

Authenticated public directory Assume there exists some party or facility, unconditionally trusted by all users. The trusted party maintains a 'trusted' public directory with users information that should be 'physically' protected from alterations and substitutions. This protection is necessary in order to avoid that a malicious adversary could substitute a fraudulent public key corresponding to a private key that he knows in the public directory. Otherwise a malicious adversary could deceive any receiver undetectable by authenticating any fraudulent message he might wish.

Obviously, a trusted authority can keep and maintain the public directory at a *central* location, where its physical security can be guaranteed. For this situation, a user doesn't need to prove his identity to gain (remote) access to the directory. (The directory contains information that is assumed to be public knowledge.) However, the communication from the directory to the user needs to be authenticated, as a user needs to trust the integrity of the received information. This can be realized with a public authentication channel of which the trusted authority owns the secret key and all users know the public key.

Certification authority and certificates The trusted authority, which takes care of the integrity of the public directory, can instead make use of his ability to create an 'off-line' public authentication channel. This channel enables the trusted authority (certification authority) to create authenticated users information in the form of certificates. (We will not make a distinction between credentials and certificates.)

A certificate comprises information (for example, the user's public key) rendered unforgeable by a signature (validating the source and its contents) with the secret key of the certification authority (CA) that issued it. A certificate is normally supplied with a unique identifier and an expiration date.

Since the CA is trusted by all users, an 'off-line' public authentication channel is created: Each user is able to verify the authenticity of a certificate and the information it contains without involvement of the CA. Note that, under the PKS-assumption, it will be computationally infeasible to alter a certificate without destroying its validity. By similar reasoning it will be computationally infeasible to falsify certificates. Therefore, user's confidence is totally dependent on the CA keeping his private key (and other sensitive information) secret, i.e., maintaining the security of his authentication channel. Depending on the application, this may require a secret-sharing system.

A protocol based on certificates consists essentially of three phases. The first two are dominated by CA's involvement. In the third and final phase, each user is able to authenticate himself with his certificate and without interference of the CA. The three phases can be summarized as follows.

1. *Set-up*

   The CA creates a public authentication channel of which only he owns the private authentication key and all users know his public key.

2. *User-validation*

   A user identifies himself to the CA. The CA is entrusted to establish first the identity and accuracy of the associated information for each potential user to whatever degree of certainty is necessary. Each user supplies the public key of his authentication channel and keeps the private key secret. The CA certifies the user's identification information and possible user's restrictions (ID-record) by signing it with his secret authentication key. Next, the CA issues the authenticated ID-record to the user as his identifying certificate.

3. *Authentication*

   Each user can present his certificate to another user (verifier). A verifier is able to verify the user's certificate with the help of CA's public authentication channel, but without CA's involvement. Moreover, the verifier knows that only the legitimate user can authenticate messages with his private key, of which he now knows the corresponding (certified) public key. Depending on the identification protocol, this information can be used to establish an authenticated shared secret.

In Section 2, some public-key distribution schemes are described, which allow (mutual) authentication with secret key exchange based on certified public keys.

The idea of identity-based public-key cryptosystem (ID-PKS) has been proposed by Shamir [14]. In ID-PKS each user's public key is based on the user's identification information, such as his name, address, etc. With this type of PKS there is no need to exchange a public key between two users. Therefore an ID-PKS does 'not' require a certified public directory. Shamir's idea can be applied to PKDS to obtain an identity-based public-key distribution system (ID-PKDS). In Section 3, examples are given of public-key distribution schemes that allow (mutual) authentication with secret key exchange based on identification information.

As mentioned before advanced identification protocols make use of so-called zero-knowledge identification protocols. In general, these schemes provide only authentication for an identity. In Section 4, a zero-knowledge identification scheme with secret key exchange is described briefly.

Finally, in Appendices A to D some supplementary information is given.

2 Public-Key Distribution

In this section some basic public-key distribution protocols are described. The protocols allow any two users to identify themselves and to establish a mutually authenticated secret. For details we refer the reader to the original papers.

2.1 The Diffie-Hellman protocol

In 1976, Whitfield Diffie and Martin Hellman [4] proposed a public-key distribution protocol based on the discrete logarithm problem. (See also Appendix A.) In elementary form, this protocol yields no authentication.

Protocol Suppose two users A and B want to share a secret S. It is assumed that the users know α (primitive element) and p (large prime number).

1. *User A randomly selects an integer S_A $(1 \leq S_A \leq p-1)$ that he keeps secret. User A then computes*

$$P_A \equiv \alpha^{S_A} \bmod p,$$

and sends P_A to B.

2. *Similarly, user B randomly selects an integer S_B $(1 \leq S_B \leq p-1)$ that he keeps secret. User B then computes*

$$P_B \equiv \alpha^{S_B} \bmod p,$$

and sends P_B to A.

3. *Both users A and B are able to compute the secret S. User A obtains S by letting*

$$S = S_{AB} \equiv P_B^{S_A} \equiv \alpha^{S_A S_B} \bmod p,$$

and user B obtains S in a similar fashion

$$S = S_{BA} \equiv P_A^{S_B} \equiv \alpha^{S_B S_A} \bmod p.$$

Note, the shared secret S depends on S_A as well as on S_B, and is not predictable by users A and B on forehand. In other words, they do not exchange a secret, but construct one. The secret S can be used, for example, as secret key in a private-key cryptosystem to enable a secure communication.

The above protocol is only resistant against a so-called passive attack (e.g., eavesdropping), because the authentication of the communication channel is not guaranteed. For example, a malicious adversary C who is able to break in on the communication channel has the opportunity to impersonate himself as user A to B and as user B to A. In this way, he will share a secret S_{AC} with user A and a secret S_{CB} with user B. However, the users A and B are under the supposition to share the same secret $S = S_{AB} = S_{BA}$. If A uses S_{AC} and B uses S_{BC} as secret key in their communication, then all traffic (messages) between A and B is legible by C. Note, the DH-protocol can be conceived as a protocol that eliminates the use of a *secrecy* channel to establish a secret S, but does not eliminate the need for an *authentication* channel.

The problem can be solved by placing the public keys P_A and P_B in a certified public directory accessible by all users. If user A wants to share a secret with another user B, then he fetches P_B from the public directory and calculates

$$S \equiv P_B^{S_A} \equiv \alpha^{S_A S_B} \bmod p.$$

In a similar manner, B obtains A's public key and computes

$$S \equiv P_A^{S_B} \equiv \alpha^{S_B S_A} \bmod p.$$

In that way, the secret S is authenticated by means of the certified public directory presumed accessible to all users. A disadvantage of this protocol is that two users A and B always share the same secret S.

The security of the DH-protocol is based on the conjecture that no method is known, given P_A and P_B, to recover either the secret integers S_A and S_B or the shared secret S, without solving the discrete logarithm problem. As stated in Appendix A, this problem is hard to solve.

In their original paper [4] Diffie and Hellman propose to use a multiplicative group $\mathbb{Z}_p^*$ modulo a prime p. Nevertheless, the DH-protocol can make use of an arbitrary group as long as the discrete logarithm problem remains hard to solve, i.e., as long as the *one-way* property is preserved.

2.2 PKDS based on the ElGamal scheme

Agnew, Mullin en Vanstone [1] proposed a public-key distribution protocol, which allows mutual authentication based on the ElGamal scheme [5]. The protocol makes use of a CA that acts like a notary service. The CA certifies public keys with an ElGamal signature (see Appendix C). The public key of the CA is known to all users and is given by

$$P_{CA} \equiv \alpha^{S_{CA}} \bmod p,$$

where S_{CA} is the private key that the CA keeps secret.

Each user U chooses at random a private key S_U $(1 \le S_U \le p - 1)$ that he keeps secret. Next, he calculates his public key P_U as follows

$$P_U \equiv \alpha^{-S_U} \bmod p.$$

In order to obtain a certified public key, user U identifies himself to the CA and presents his public key P_U. The CA signs the public key P_U and issues a certificate C_U.

The signing procedure is as follows. The CA chooses randomly an integer W_U relatively prime to $p - 1$ and computes an integer X_U such that

$$P_U \equiv S_{CA} \times \alpha^{W_U} + W_U X_U \bmod (p - 1).$$

The (ElGamal-)signature for P_U is the pair (W_U, X_U). The CA gives the certificate $C_U = (W_U, X_U)$ to user U.

Henceforth, each user U possesses a public key P_U, together with a secret key S_U and a corresponding certificate $C_U = (W_U, X_U)$. All users know α, p and the CA's public key P_{CA}. The protocol runs as follows.

Protocol Suppose two users A and B require to share a secret S.

1. *User A obtains the pair (P_B, C_B) from user B and verifies the public key P_B as follows*

$$\alpha^{P_B} \overset{?}{\equiv} (P_{CA})^{\alpha^{W_B}} \times (\alpha^{W_B})^{X_B} \bmod p.$$

2. *If the verification passes, user A applies the ElGamal-protocol (Appendix B) to exchange a message K_A. For this purpose he randomly selects two integers K_A and R_A and computes*

$$\alpha^{R_A} \bmod p \quad and \quad P_B^{R_A} K_A \bmod p,$$

which are forwarded to user B.

3. *Upon receipt, user B recovers the message K_A as follows*

$$K_A \equiv (\alpha^{R_A})^{S_B} \times (P_B^{R_A} K_A) \equiv \alpha^{R_A S_B} \alpha^{-S_B R_A} \times K_A \bmod p.$$

4. *For mutual authentication user B receives (P_A, C_A) and verifies the public key P_A. After positive verification he calculates the secret S*

$$S = S_{BA} \equiv P_A^{S_B \times K_A} \equiv \alpha^{-S_B S_A} \times \alpha^{K_A} \bmod p.$$

Similarly, A computes S by letting

$$S = S_{AB} \equiv P_B^{S_A \times K_A} \equiv \alpha^{-S_A S_B} \times \alpha^{K_A} \bmod p.$$

A disadvantage of this protocol is the lack of symmetry in the shared secret S. The first part of the secret S depends entirely on S_A and S_B, and remains the same for two users (like in the DH-protocol). The randomly generated K_A is the only part of S that will be different each time the protocol is initiated. Therefore, only user A has influence on the value of S. However, in non-interactive situations (e.g., Email) the ability to create a secret without interaction with user B is necessary. Here a lack of symmetry seems unavoidable.

For details, see [1].

2.3   Composite Diffie-Hellman protocol

As stated in Section 2.1, the DH-protocol can be generalized to an arbitrary group as long as the one-way property is preserved. Instead of using a multiplicative group $\mathbb{Z}_p^*$ in which p is a large prime, McCurley [10] and Shmuely [15] proposed a variation using the multiplicative group $\mathbb{Z}_m^*$ with composite m. Factoring large composite numbers is known to be at most as hard as computing discrete logarithms modulo a composite number. (The factoring problem can be reduced to the general discrete logarithm problem.) McCurley proved that if the integers r and s in the factors $p = 8r + 3$ and $q = 8s - 1$ of the modulus $m = pq$ satisfy

$$2r + 1 \text{ has a large prime factor,} \quad 4r + 1 \text{ and } 8r + 3 \text{ are prime numbers,}$$

$$s \text{ has a large prime factor,} \quad 4s - 1 \quad \text{and } 8s - 1 \quad \text{are prime numbers,}$$

then any algorithm that will break the composite DH-protocol (CDH) for base $\alpha = 16$ can be used to factor m, and thus break the original DH-protocol modulo the prime factors p and q. Provided m is hard to factor, this gives evidence that the CDH-protocol is hard to break even for a fixed fraction of instances.

Protocol   Suppose two users A and B request to share a secret S. It is assumed that they know the composite number m and the base $\alpha = 16$.

1. *User A randomly selects an integer S_A ($1 \leq S_A \leq m-1$) that he keeps as his (secret) private key. User A computes the public key $P_A \equiv \alpha^{S_A} \bmod m$, and sends the result to user B.*

2. *Similarly, user B randomly selects an integer S_B ($1 \leq S_B \leq m - 1$) that he keeps as his (secret) private key. User B computes the public key $P_B \equiv \alpha^{S_B} \bmod m$, and sends the result to user A.*

3. *After receiving P_A and P_B, both users are able to compute the secret S as follows*

$$S = S_{AB} \equiv P_A^{S_B} \equiv \alpha^{S_A S_B} \bmod m;$$

$$S = S_{BA} \equiv P_B^{S_A} \equiv \alpha^{S_B S_A} \bmod m.$$

McCurley [10] remarks that the specific choice of p, q and α are by no means the only ones possible for which it is possible to prove the direct link between the DH-protocol and factoring. Details can be found in [10].

2.4   Authenticated CDH-protocol with random secret

A disadvantage of the original DH-protocol with authenticated public keys is that two users A and B always share the same secret $S = S_{AB} = S_{BA}$. The secret S depends solely on the secret keys S_A and S_B. This undesirable property is eliminated in a protocol proposed by Yacobi and Shmuely [18]. Their basic protocol makes use of the CDH-protocol, i.e., the DH-protocol with composite modulus.

Each user U possesses a certified public key P_U, which is computed as follows

$$P_U \equiv \alpha^{S_U} \bmod m,$$

where the private key S_U is kept secret. The private key S_U can be considered as a private key of a public authentication channel of which P_U is the public key. The protocol runs as follows.

Protocol Suppose two users A and B want to share a secret S. It is assumed that they know the composite number m and the base α.

1. *User A chooses at random an integer R_A ($0 \le R_A \le m - 1$) which he keeps secret. He further computes the integer $X_A = R_A + S_A$ ($0 \le X_A \le 2m - 2$), and sends the result to user B.*

2. *Similarly, user B chooses at random an integer R_B ($0 \le R_B \le m - 1$) which he keeps secret. He further computes the integer $X_B = R_B + S_B$ ($0 \le X_B \le 2m - 2$), and sends the result to user A.*

3. *After receiving X_B, user A is able to calculate the secret S*

$$S = S_{AB} \equiv (\alpha^{X_B} P_B^{-1})^{R_A} \equiv (\alpha^{R_B + S_B} \alpha^{-S_B})^{R_A} \equiv \alpha^{R_A R_B} \bmod m.$$

In the same way, after receiving X_A, user B can compute S by letting

$$S = S_{BA} \equiv (\alpha^{X_A} P_A^{-1})^{R_B} \equiv (\alpha^{R_A + S_A} \alpha^{-S_A})^{R_B} \equiv \alpha^{R_B R_A} \bmod m.$$

The value of the shared secret S depends entirely on R_A and R_B, which are generated at random whenever the protocol is initiated. As a result, the value of S will be different each time. The authentication takes place with public keys. To complete the authentication it is necessary that both users are convinced that they share the same secret S. A general solution is the ability to synchronize their secret communication, i.e., by encrypting redundant messages using S.

Although the protocol clearly leaks information about S_A and S_B, Yacobi and Shmuely prove the security of their protocol with the amortized security [6]. For details see [18].

2.5 Authenticated DH-protocol with random secret

A disadvantage of the previous CHD-protocol (Section 2.4) is that we do not know the factorization of m, i.e., we cannot reduce the integer $X_U = R_U + S_U$ modulo $\phi(m) = (p - 1)(q - 1)$. (Note that if we know the value of $\phi(m)$ then we are able to factor m.) Hence, subsequent execution of this protocol leaks information about S_U. To avoid this leakage, a prime modulus p can be used in stead of the composite modulus m. Besides an extra modular reduction, the protocol remains the same.

Each user U possesses a certified public key P_U, which is computed as follows

$$P_U \equiv \alpha^{S_U} \bmod p,$$

where the private key S_U is kept secret. The private key S_U can be considered as a private key of a public authentication channel of which P_U is the public key. The protocol runs as follows.

Protocol Suppose two users A and B want to share a secret S. It is assumed that they know the prime number p and the base α.

1. *User A chooses at random an integer R_A $(0 \leq R_A < p - 1)$ which he keeps secret. He further computes the integer $X_A = R_A + S_A \bmod p - 1$ and sends the result to user B.*

2. *Similarly, user B chooses at random an integer R_B which he keeps secret. He further computes the integer $X_B = R_B + S_B \bmod p - 1$ and sends the result to user A.*

3. *After receiving X_B, user A is able to calculate the secret S*

$$S = S_{AB} \equiv (\alpha^{X_B} P_B^{-1})^{R_A} \equiv (\alpha^{R_B + S_B} \alpha^{-S_B})^{R_A} \equiv \alpha^{R_A R_B} \bmod p.$$

In the same way, after receiving X_A, user B can compute S by letting

$$S = S_{BA} \equiv (\alpha^{X_A} P_A^{-1})^{R_B} \equiv (\alpha^{R_A + S_A} \alpha^{-S_A})^{R_B} \equiv \alpha^{R_B R_A} \bmod p.$$

As in the previous CHD-protocol (Section 2.4), the value of the shared secret S depends entirely on R_A and R_B and the authentication is indirectly.

3 Identity-based PKDS

The idea of identity-based public-key cryptosystem (ID-PKS) has been proposed by Shamir [14]. In ID-PKS each user's public key is based on his own identification information, such as his name, address, etc. With this type of PKS there is no need to exchange a public key between two users. The need for a CA is to 'publish' the identification information of a new user. The CA is not required to maintain a list of users. Shamir's idea can be applied to PKDS to obtain an identity-based public-key distribution system (ID-PKDS). In this section, two such ID-PKDS are discussed in which the mutual authentication is based on user's identification information.

3.1 Tanaka-Okamoto's ID-PKDS

Tanaka and Okamoto (e.g., [13, 17]) have proposed several identity-based protocols, which strictly speaking are based on the same idea. Two versions are described, which make use of a trusted authority CA to create certified keys based on ID-information.

3.1.1 First Protocol

The CA generates a RSA modulus $m(= pq)$, an encryption exponent E_{CA} and a decryption exponent D_{CA}, satisfying

$$E_{CA} D_{CA} \equiv 1 \bmod (p-1)(q-1).$$

The CA keeps D_{CA} secret, but makes m and E_{CA} publicly known. Each user U has a unique (registered) ID-string: $ID_U = (name, address, \ldots)$. The CA generates for each user U the secret information S_U, using the registered identity information ID_U, as follows

$$S_U \equiv (ID_U)^{-D_{CA}} \bmod m, \quad \text{so that} \quad ID_U \equiv S_U^{-E_{CA}} \bmod m.$$

Protocol 1 Suppose two users A and B request to share a secret S. It is assumed that they know the registered ID-strings, the modulus m and the encryption exponent E_{CA} of the CA.

1. *User A chooses at random an integer R_A and computes*

$$P_A \equiv S_A \alpha^{R_A} \bmod m,$$

where S_A and R_A are kept secret. The result P_A is sent to user B. In a similar fashion, user B computes P_B and forwards it to user A.

2. *Upon receipt of P_A, user B is able to compute the secret S*

$$S = S_{BA} \equiv (P_A^{E_{CA}} \times ID_A)^{R_B} \equiv \alpha^{E_{CA}} \times \alpha^{R_B R_A} \bmod m.$$

Similarly, A receives P_B and obtains the secret as follows

$$S = S_{AB} \equiv (P_B^{E_{CA}} \times ID_B)^{R_A} \equiv \alpha^{E_{CA}} \times \alpha^{R_A R_B} \bmod m.$$

3.1.2 Second Protocol

The CA generates a RSA modulus $m(= pq)$ and an integer E_{CA} relatively prime with $(p-1)(q-1)$. Each user U has a unique (registered) ID-string $ID_U = (name, address, \ldots)$. The CA generates for each user U the secret information S_U as follows

$$S_U^{E_{CA}} \equiv (ID_U)^{-1} \bmod m, \quad \text{so that} \quad ID_U \equiv S_U^{-E_{CA}} \bmod m.$$

The CA secretly distributes the secret information S_U to each user. Upon receipt, each user computes his public key

$$P_U \equiv S_U \alpha^{R_U} \bmod m,$$

where R_U is randomly generated and kept secret along with S_U. Each user U submits P_U to a public directory. The protocol runs as follows.

Protocol 2 Suppose two users A and B wish to share a secret S. It is assumed that they know the registered ID-strings, the modulus m and the encryption exponent E_{CA} of the CA.

1. *User A randomly chooses an integer K_A and computes*

$$Y_A \equiv S_A \alpha^{K_A} \bmod m.$$

A keeps K_A secret and sends the result Y_A to user B.
Next, he computes the secret key

$$S = S_{AB} \equiv (P_B^{E_{CA}} ID_B)^{K_A} \equiv \alpha^{R_B E_{CA}} \alpha^{K_A} \bmod m.$$

2. *Upon receipt of Y_A, user B is able to compute the secret as well*

$$S = S_{BA} \equiv (Y_A^{E_{CA}} \times ID_A)^{R_B} \equiv (S_A^{E_{CA}} ID_A \alpha^{E_{CA} K_A})^{R_B} \equiv \alpha^{E_{CA} R_B} \alpha^{K_A} \bmod m.$$

The first part of the secret S depends entirely on E_{CA} and R_B, and remains the same for two users (like in the DH-protocol). The randomly generated K_A is the only part of S that will be different each time the protocol is executed. Therefore, only user A has influence on the value of S. This is a desired property in non-interactive situations (e.g., Email), i.e., the ability of user A to create a secret without interaction of user B. In the above protocol, S_A is used to send and R_B to receive messages.

3.2 Günther's ID-protocol

Günther's ID-PKDS is based on a combination of ElGamal's scheme and the Diffie-Hellman protocol. It makes use of a CA which is in possession of a publicly known one-way function $f(.)$ and a public key P_{CA} satisfying

$$P_{CA} \equiv \alpha^{S_{CA}} \bmod p,$$

where S_{CA} is kept secret.

Each user U has a unique ID-string: $ID_U = (name, address, \ldots)$, which he presents to the CA. The CA computes a verification identity

$$VID_U = f(ID_U).$$

Hereafter, the CA certifies the VID_U with an ElGamal signature (see Appendix C). The CA generates at random an integer K and calculates $R_U \equiv \alpha^K \bmod p$. Next, he solves X_U from

$$VID_U \equiv S_{CA}R_U + KX_U \bmod (p-1).$$

The protocols runs as follows.

Protocol Suppose two users A and B request to share a secret S. Each user U has a unique ID-string, a certificate $C_U = (VID_U, R_U, X_U)$ and knows CA's public information: $f(.)$, p, α and P_{CA}.

1. User A receives user's B ID-string ID_B and R_B and computes $VID_B = f(ID_B)$. Next, A verifies as follows

$$R_B^{X_B} \stackrel{?}{\equiv} \alpha^{VID_B} P_{CA}^{-R_B} \bmod p.$$

In a similar fashion, user B verifies the information he receives from user A.

2. After positive verification, user A generates two integers T_A and W_A at random, computes

$$U_A \equiv R_B^{T_A} \bmod p \quad and \quad V_A \equiv \alpha^{W_A} \bmod p,$$

and sends the results to user B. In a similar way, user B obtains

$$U_B \equiv R_A^{T_B} \bmod p \quad and \quad V_B \equiv \alpha^{W_B} \bmod p,$$

and sends these results to user A.

3. User A computes the secret $S = S_{AB}$ as follows

$$S = S_{AB} \equiv (R_B^{X_B})^{T_A} \times U_B^{X_A} V_B^{W_A} \equiv R_B^{X_B T_A} R_A^{X_A T_B} \times \alpha^{W_A W_B} \bmod p.$$

In the same way, user B computes the secret $S = S_{BA}$

$$S = S_{BA} \equiv (R_A^{X_A})^{T_B} \times U_A^{X_B} V_A^{W_B} \equiv R_A^{X_A T_B} R_B^{X_B T_A} \times \alpha^{W_A W_B} \bmod p.$$

The last term $\alpha^{W_a W_b}$ resembles the original Diffie-Hellman protocol. If A and B are not impersonated, then finding the secret S is as difficult as breaking the DH-protocol. If CA's private information is compromised, then the secrecy of previously generated secrets S is not affected, only the authenticity of future secrets.

Details can be found in [7]

4 Zero-Knowledge Authentication with Secret Exchange

In [2] a framework for zero-knowledge key distribution with authentication is proposed. The protocol is based on an interactive proof system for functions, and combines a zero-knowledge identification scheme with a method of exchanging a secret. In this proof system the prover demonstrates his ability to compute the inverse of a given function.

Suppose, the protocol can be simulated by any verifier (honest or dishonest) in reasonable time without interaction of the prover. Then the simulation surely doesn't convey any knowledge about the prover's private information. In a zero-knowledge scheme, a simulation and the genuine conversation are indistinguishable. Hence, the real conversation will not convey knowledge of the private key either (in contrast to the protocol in Appendix D). The zero-knowledge property ensures the prover that he doesn't give away any information, which makes it easier for any opponent to cheat.

In this section, a brief outline of the protocol is presented based on the Rabin variant of the RSA scheme. (Note that for RSA it is not yet known that computing the inverse of the public key is equivalent to computing the private key.)

Let $p_U \equiv 3 \bmod 8$ and $q_U \equiv 7 \bmod 8$ be the secret prime factors of m_U. Define $d_U = [(p_U - 1)(q_U - 1) + 4]/8$. Each user U randomly selects his private decryption function $D_U(X) \equiv X^{d_U} \bmod m_U$ and his corresponding public encryption function $P_U(X) \equiv X^2 \bmod m_U$, where m_U is an L bit modulus. A sequence of n bits $x_0, \ldots, x_{n-1}$ is interpreted as a nonnegative integer X such that

$$X = \sum_{i=0}^{n-1} x_i 2^{n-i-1},$$

i.e., the more significant bits being first in the sequence. Let $[X]_{a,b} = [x_0, \ldots, x_{n-1}]_{a,b}$ denote the bits $x_a, x_{a+1}, \ldots, x_b$ of X.

The protocol runs as follows.

Protocol Suppose two users A and B wish to establish an authenticated secret S of length l.

- *User B chooses at random an integer S_B and and computes*

$$P_A(S_B) = R_B \quad and \quad P_A(R_B) = C_B.$$

User B sends the results C_B and $V_B = [R_B]_{L-64,L-1}$ to A.

- *Upon receipt, A decrypts C_B and obtains*

$$\hat{R}_B = D_A(C_B).$$

If $[\hat{R}_B]_{L-64,L-1} = V_B$ then let $W_B = [\hat{R}_B]_{L-128,L-65}$, else if $[m_A - \hat{R}_B]_{L-64,L-1} = V_B$ then let $W_B = [m_A - \hat{R}_B]_{L-128,L-65}$. When the last check also fails the execution of the protocol stops, if not A returns W_B.

- *Upon receipt, B compares W_B with $[R_B]_{L-64,L-1}$. If equality holds, B is convinced of A's identity. Otherwise he rejects A's proof.*

- *Finally, both users compute their l bit secret key S. User B by letting*

$$S = S_{BA} = [R_B]_{L-128-l,L-129},$$

and user A by letting

$$S = S_{AB} = [\hat{R}_B]_{L-128-l,L-129} \quad \text{if} \quad V_B = [\hat{R}_B]_{L-64,L-1},$$

or

$$S = S_{AB} = [m_A - \hat{R}_B]_{L-128-l,L-129} \quad \text{if} \quad V_B = [m_A - \hat{R}_B]_{L-64,L-1}.$$

User B is in two ways in advantage over user A: (1) He chooses the value of R_B and (2) he is convinced of A's identity. This in contrast to user A: He has no influence on the choice of R_B and he is not sure of the identity of B. As a consequence, the secret key S is authenticated only from A to B and not vice versa. As stated in [2], this can be solved in the following way. User A can prove himself to B while B is proving himself to A. At the end, each user has two keys. The bitwise exclusive or of these is used as the common secret key S.

References

[1] Agnew, G., R. Mullin and S. Vanstone, An Interactive Data Exchange Protocol Based on Discrete Exponentiation, *Advances in Cryptology - Eurocrypt'88*, Lecture Notes in Computer Science, vol. 330, Springer-Verlag, pp. 159-166, 1989.

[2] Brandt, J., I. Damgård, P. Landrock and T. Pedersen, Zero-Knowledge Authentication Scheme with Secret Key Exchange, to appear.

[3] Coppersmith, D., A. Odlyzko and R. Schroepel, Discrete logarithms in $GF(p)$, *Algorithmica*, vol. 1, pp. 1-15, 1986.

[4] Diffie, W., and M.E. Hellman, New directions in cryptography, *IEEE Trans. on Inform. Theory*, vol. IT-22, no. 6, pp. 644-654, 1976.

[5] ElGamal, T., A Public Key Cryptosystem and a Signature Scheme Based on Discrete Logarithms, *IEEE Trans. on Inform. Theory*, vol. IT-31, no. 4, pp. 469-472, 1985.

[6] Goldreich, O., and A.L. Levin, Hard-Core Predicate for All One-Way Functions, STOC'89, pp. 25-32.

[7] Günther, C.G., An Identity-Based Key-Exchange Protocol, *Advances in Cryptology - Eurocrypt'89*, Lecture Notes in Computer Science, vol. 434, Springer-Verlag, pp. 29-37, 1990.

[8] Knuth, D.E., The Art of Computer Programming, *Vol.2: Seminumerical Algorithms*, 2nd edition, Addison-Wesley, 1981.

[9] Massey, J.L., An Introduction to Contemporary Cryptology, *Proceedings of the IEEE*, vol. 76, no. 5, pp. 533-549, 1988.

[10] McCurley, K.S., A Key Distribution System Equivalent to Factoring, *Journal of Cryptology*, vol. 1, pp. 95-105, 1988.

[11] McCurley, K.S., The discrete logarithm problem, *Cryptography and Computational Number Theory*, Proc. Symp. Appl. Math., Amercan Mathematical Society, 1990.

[12] Odlyzko, A., Discrete logarithms in finite fields and their cryptographic significance, *Advances in Cryptology*, Proceedings of Eurocrypt'84, Lecture Notes in Computer Science, vol. 209, Springer-Verlag, pp. 224-314, 1985.

[13] Okamoto, E., and K. Tanaka, Key Distribution System Based on Identification Information, *IEEE Journal on selected areas in communications*, vol. 7, no. 4, 1989.

[14] Shamir, A., Identity-based cryptosystems and signatures schemes, *Advances in Cryptology - Crypto'84*, Lecture Notes in Computer Science, vol. 209, Springer-Verlag, pp. 47-53, 1985.

[15] Shmuely, Z., Composite Diffie-Hellman Public-Key Generating Systems Are Hard To Break, Technical Report 356, Computer Science Department, Technion-Israel Institute of Technology, 1985.

[16] Simmons, G.J., A Survey of Information Authentication, *Proceedings of the IEEE*, vol. 76, no. 5, pp. 603-620, 1988.

[17] Tanaka, K., and E. Okamoto, Key Distribution System for Mail Systems Using ID-related Information Directory, *Computer & Security*, vol. 10, pp. 25-33, 1991.

[18] Yacobi, Y., and Z. Shmuely, On Key Distribution Systems, *Advances in Cryptology - Crypto'89*, Lecture Notes in Computer Science, vol. 435, Springer-Verlag, pp. 344-355, 1990.

A The Discrete Logarithm Problem

In cryptography, many protocols are based on so called *one-way functions*, i.e., functions that are *easy* to calculate but *hard* to invert. Although it has never been proved that such functions really exist, the *modular* or *discrete exponential function* is usually assumed to be one-way. The discrete exponential function is defined as

$$f(x) \equiv \alpha^x \bmod p,$$

where p is a large prime and x, α are integers between 1 and $p - 1$. Moreover, α is a so called primitive element, that is, $\alpha, \alpha^2, \ldots, \alpha^{p-1} \bmod p$ take all values from $\{1, 2, \ldots, p-1\}$ in some order. The modular exponentiation is relatively easy to calculate. The opposite problem of finding x from $f(x)$ is called *the discrete logarithm problem*. The discrete logarithm problem is considered to be hard to solve when $p - 1$ has large prime factors. Ideally, the prime p should be of the form $2q + 1$ with q prime. To date, there is no proof that the discrete logarithm problem is really hard to solve. However, it is known that certain bits of x (for example, the least significant bit) can be retrieved easily. For more details we refer to [3, 11] and [12].

B ElGamal's Public-Key Cryptosystem

ElGamal described in [5] a public key cryptosystem based on the discrete logarithm problem. Each user A chooses a private key S_A ($1 \leq S_A \leq p - 1$) and calculates its public key as follows

$$P_A \equiv \alpha^{S_A} \bmod p,$$

and places P_A in a certified public directory. If A wants to send a message m ($0 \leq m \leq p - 1$) to B, then he randomly chooses an integer R_A ($1 \leq R_A \leq p - 1$) and send to B the pair

$$c = (c_1, c_2) \equiv (m P_B^{R_A}, \alpha^{R_A}) \bmod p.$$

User B decrypts c as follows

$$m \equiv c_1 P_B^{-R_A} \bmod p,$$

where $y_B^{-R_A}$ follows from

$$c_2^{p-1-S_B} \equiv (\alpha^{R_A})^{p-1-S_B} \equiv (\alpha^{R_A})^{-S_B} \equiv (\alpha^{S_B})^{-R_A} \equiv P_B^{-R_A} \bmod p.$$

More details can be found in [5].

C ElGamal's Digital Signature Scheme

The most well known digital signature scheme using discrete exponentiation is due to ElGamal [5]. The public key for user A is given by

$$P_A \equiv \alpha^{S_A} \bmod p,$$

where S_A is his private key.

To sign a message m, user A applies his private key S_A to create a signature for m in such a way that all users are able to verify the authenticity of the signature using the certified public key P_A of user A. It is assumed that no one can forge a signature without knowing the private key S_A.

The *signing procedure* is as follows. Let m ($0 \leq m \leq p - 1$) be the message to be signed. User A randomly chooses an integer K ($1 \leq K \leq p - 1$) such that K and $p - 1$ are relatively prime and computes

$$R_A \equiv \alpha^K \bmod p.$$

Hereafter, A calculates an integer T_A such that

$$m \equiv S_A R_A + K T_A \bmod (p-1).$$

The signature for m is the pair (R_A, T_A).

The *verification procedure* consists of verifying the equality

$$\alpha^m \stackrel{?}{\equiv} P_A^{R_A} R_A^{T_A} \bmod p.$$

More details can be found in [5].

D Proof of Identity

The following protocol is due to J. Omura. User A randomly chooses an integer S_A ($1 \leq S_A \leq p-1$) that he keeps secret as his private key. Next, user A computes his public key

$$P_A \equiv \alpha^{S_A} \bmod p,$$

which he places in a certified public directory. Suppose user B (verifier) wants to verify the identity of user A (prover). He proceeds as follows. Verifier B chooses a random number S_B ($1 \leq S_B \leq p-1$) and sends to A a challenge

$$C_B \equiv \alpha^{S_B} \bmod p.$$

User A computes

$$Z \equiv C_B^{S_A} \equiv \alpha^{S_A S_B} \bmod p,$$

and returns the result to V. The verifier computes

$$P_A^{S_A} \equiv \alpha^{S_A S_B} \bmod p,$$

and compares the result with Z.

If verifier B cheats and sends S_B instead of α^{S_B}, then B obtains $S_B^{S_A}$ from A, which he is not able to compute himself. Therefore this protocol is not a zero-knowledge proof of identity.

Information Authentication: Hash Functions and Digital Signatures

Bart Preneel*, René Govaerts, Joos Vandewalle

ESAT-COSIC Laboratory, K.U.Leuven
K. Mercierlaan 94, B-3001 Leuven, Belgium

bart.preneel@esat.kuleuven.ac.be

Abstract. The goal of this paper is to discuss techniques for the protection of the authenticity of information. The theoretical background is sketched, but most attention is paid to overview the large number of practical constructions for symmetric authentication and digital signatures.

1 Introduction

In former days, the protection of information was mainly an issue of physical security and selection and motivation of people. To keep information confidential, it was carefully locked in a vault and the discretion of the personnel prevented data from being disclosed. The protection of authenticity relied exclusively on the impossibility of forging certain documents (as is now still the case e.g., for visa) and of manual signatures. The identification of people relied on eye-to-eye contact and transfer of important information was done by a courier.

The first evolution was the processing of information stored on punch cards or tapes with digital computers. The processing capabilities of the computers increased and large amounts of data are now stored on magnetic or optical carriers. Transfer of information is realized through both local and worldwide telecommunication networks. Information processed in computers and under transfer in communication networks is certainly more vulnerable to both passive and active attacks. If messages are exchanged under electronic form, it is not sufficient to write the name of the sender at the end. Everyone can modify the message or change sender or recipient. This type of threat is especially important in the context of financial transactions. Information stored in computers can also easily be modified without leaving any trace. Computers are vulnerable to more specific threats: computer viruses, Trojan horses, worms, and logic bombs can cause considerable damage. Apart from outsiders breaking into computer systems, the people operating the computer can also be malicious. Copying or changing electronic data is much easier for them than modifying data stored on paper. Another application that will become more and more important is the

* NFWO aspirant navorser, sponsored by the National Fund for Scientific Research (Belgium).

protection of the authenticity of pictures and moving images (e.g., videoconferencing). As one can expect that it will become feasible to "edit" moving pictures and make a person say and do things he or she never said or did, it is required that one can guarantee the authenticity of moving images. This will impose high requirements on the throughput. Other applications where authentication is important are alarm systems, satellite control systems, distributed control systems, and systems for access control.

The intention of message authentication is to protect the communicating parties against attacks of a third party. However, a different threat emerges when the two communicating parties are mutually distrustful and try to perform a *repudiation*. This means that sender or receiver will try to modify a message and/or deny to have sent or received a particular message. In paper documents, protection against this type of attack is offered by a handwritten signature. It is clear that in case of electronic messages, a simple name at the end of the message offers no protection at all. This is analogous to the fact that a photocopy of a document with a manual signature has no value, as one can always produce a bogus document with cut and paste operations. Similarly, a manual signature under a message sent through a telefax or via a computer network has no value. A typical example of this fraud is the electronic communication of orders to a stockbroker. The customer can place an order to buy shares of company X. If some days later the transaction turns out badly, he will try to deny his order. If on the other hand, the transaction is successful, the stockbroker might claim that he never received the order with the intention to keep the profit. In case of a dispute, a third party (e.g., a judge) has to take a decision.

The goal of this paper is not to enumerate all threats, but to discuss solutions. A very important observation is that cryptographic techniques are essential for securing information in computer systems and networks. In case of protection against a third party, the term *symmetric authentication* is used, and the cryptographic techniques employed are cryptographic hash functions. In case of protection against repudiation, one uses the term *asymmetric authentication* and the corresponding cryptographic techniques are digital signatures.

Of course, cryptographic techniques are not sufficient. Other aspects, e.g., management and selection of personnel are also very important. We can use here the old cliché saying that a chain is as strong as its weakest link. Moreover, existing solutions will never be completely sufficient and therefore audit trails and arbitration procedures are necessary.

In a first part of this overview paper, the difference between symmetric authentication algorithms and digital signatures is explained in more detail, and the need for both techniques is demonstrated. Three approaches to solve cryptographic problems are compared. Most attention is paid here to unconditionally secure authentication. The second part deals with cryptographic hash functions. First definitions are given, and subsequently applications are discussed. Then an extensive overview is given of proposed schemes together with a discussion of their performance and a taxonomy for methods of attack. In a third part a brief overview is given of practical digital signature schemes.

2 Information Authentication

2.1 Symmetric Authentication

The protection of the authenticity of information includes two aspects:

- the protection of the originator of the information, or in ISO terminology [71] data origin authentication.
- the fact that the information has not been modified, or in ISO terminology [71] the integrity of the information.

The first aspect can be present in a hidden way, e.g., in case the information is read from the hard disk of a personal computer, and one implicitly trusts the source of the information. Other aspects that can be important are the timeliness, the sequence with respect to other messages, and the destination.

Until recently, it was generally believed that encryption of information suffices to protect its authenticity. The argument was that if a ciphertext resulted after decryption in meaningful information, it should be originated with someone who knows the secret key, guaranteeing authenticity of message and sender. Two counterexamples will show that this belief is wrong: the protection of integrity is dependent on the encryption algorithm and on the mode in which the algorithm is used.

The Vernam cipher [143], where a random key string is added modulo 2 to the plaintext, offers unconditional secrecy, but an active attacker can change any bit of the plaintext by simply flipping the corresponding bit of the ciphertext. This observation is also valid for any additive stream cipher and for the Output FeedBack (OFB) mode of any block cipher. It holds partially if the cipher is used in Cipher FeedBack (CFB) or Cipher Block Chaining (CBC) mode [43, 52, 76].

If a plaintext longer than one block is enciphered with a block cipher in ECB mode, an active attacker can easily reorder the blocks. Another example is the vulnerability to active attacks of a plaintext encrypted in Cipher Feedback Mode (CFB). Due to the self-synchronization properties, any modification in the ciphertext will cause a corresponding modification to the plaintext and will subsequently garble the next part of the plaintext. When the error has shifted out of the feedback register, the ciphertext will be deciphered correctly again. If the last part of the ciphertext is modified, it is completely impossible to detect this. If the garbling occurs in the middle of the plaintext, it can only be detected based on redundancy, as will be discussed in the next paragraph.

In other modes (like the CBC mode) every ciphertext bit is a complex function of the previous plaintext bits and an initial value. If the modification of a single ciphertext bit results in t bits of the plaintext being garbled, the probability that the new plaintext will be accepted as meaningful equals 2^{-tD}, where D is the redundancy in the information. In case of natural language coded with 5 bits per character the redundancy per bit $D \approx 0.74$, and this probability is equal to $2^{-22.2}$ for $t = 30$. However, if $D = 0$ all messages are meaningful, and encryption offers no authentication, independently of the encryption algorithm or of the mode. This means that an attacker can modify messages or forge messages of his choice. The limitation is that he does not know on beforehand what the corresponding plaintext will be, but many applications can be considered where such an attack would cause serious problems. Note that even if redundancy is present, a human checker or a designated computer program is required to check its presence.

In order to protect the integrity, special redundancy will have to be added, and if the information is to be linked with an originator, a secret key has to be involved in the process (this assumes a coupling between the person and his key), or a separate integrity channel has to be provided. Hence two basic methods can be identified.

— The first approach is analogous to the approach of a symmetric cipher, where the secrecy of large data quantities is based on the secrecy and authenticity of a short key. In this case the authentication of the information will also rely on the secrecy and authenticity of a key. To achieve this goal, the information is compressed to a quantity of fixed length, which is called a *hashcode*. Subsequently the hashcode is appended to the information. The function that performs this compression operation is called a *hash function*. The basic idea of the protection of the integrity is to *add redundancy* to the information. The presence of this redundancy allows the receiver to make the distinction between authentic information and bogus information.

  In order to guarantee the origin of the data, a secret key that can be associated to the origin has to intervene in the process. The secret key can be involved in the compression process or can be used to protect the hashcode and/or the information. In the first case the hashcode is called a Message Authentication Code or MAC, while in the latter case the hashcode is called a Manipulation Detection Code or MDC.

— The second approach consists of basing the authenticity (both integrity and origin authentication) of the information on the authenticity of a Manipulation Detection Code or MDC. A typical example for this approach is a computer user who will calculate an MDC for all its important files. He can store this collection of MDC's on a floppy, that is locked in his safe, or he can write them down on a piece of paper. If he has to transfer the files to a remote friend, he can simply send the files and communicate the MDC's via telephone. The authenticity of the telephone channel is offered here by voice identification.

The addition of redundancy is certainly not sufficient. Special care has to be taken against high level attacks, like a replay of an authenticated message (cf. Sect. 8.5).

Both approaches do not work if sender and receiver do not trust each other. In the first approach, they share the same secret key. If one of the parties claims that the information was changed by the other party, a judge can not make a distinction between them, even if they disclose their common secret key. The second approach can only provide non-repudiation if both parties trust the authenticity of the MDC: in practice however this is difficult to realize, as both parties have a similar access to that channel.

2.2 Asymmetric Authentication

If one wants to be protected against insider attacks, one needs an electronic equivalent of a manual signature. In this case a third party will be able to distinguish between both parties, based on the fact that the capabilities of both parties are different. The concept of a digital signature was introduced by W. Diffie and M. Hellman in [46]. The requirements are that the signature depends on the information to be signed (since it is not physically connected to the document) and that the signer is the only person who can produce a signature (this means no one else can forge a signature, which implies that the signer can not deny that he has produced it). A digital signature scheme consists of the following elements:

- an initialization phase (e.g., key generation and general setup),
- a signing process, where the signature is produced,
- a verification process, where the receiver (or a judge) verifies whether the signature is correct.

Digital signatures in this sense can be produced based on tamper resistant devices, conventional one-way functions, or public-key techniques. In Sect. 9 a brief overview will be given of practical signature schemes.

Several generalizations have been defined, with more players in the game and with different notions of security. Examples of such extensions are the following: *arbitrated signatures*, where the signing and verification process involves interaction with a third party, *group signatures*, where signers and/or verifies are members of a group, *blind signatures*, where the signer signs a 'blinded' or 'masked' message, and *undeniable* or *invisible signatures*, where the signature can only be verified with the cooperation of the signer.

For a more extensive treatment of the general aspects of digital signature schemes the reader is referred to [102].

3 Three Approaches in Cryptography

In present day cryptography, three approaches can be identified to solve most problems comprising information secrecy and information authenticity. These approaches differ in the assumptions about the capabilities of an opponent, in the definition of a cryptanalytic success, and in the notion of security. Our taxonomy deviates from the approach by G. Simmons [139], and is based on the taxonomy for stream ciphers of R. Rueppel [133]. A first method is based on information theory, and it offers unconditional security, i.e., security independent of the computing power of an adversary. The complexity theoretic approach starts from an abstract model for computation, and assumes that the opponent has limited computing power. The system based approach tries to produce practical solutions, and the security estimates are based on the best algorithm known to break the system and on realistic estimates of the necessary computing power or dedicated hardware to carry out the algorithm. In [139] the second and third approach are lumped together as computationally secure, and in [133] a fourth approach is considered, in which the opponent has to solve a problem with a large size (namely examining a huge publicly accessible random string); it can be considered as both computationally secure and information theoretically secure.

3.1 Information Theoretic Approach

This approach results in a characterization of unconditionally secure solutions, which implies that the security of the system is independent of the computing power of the opponent. E.g., in case of privacy protection, it has been shown by C. Shannon [137] that unconditional privacy protection requires that the entropy of the key is lower bounded by the entropy of the plaintext. It should be remarked that both unconditional privacy and unconditional authenticity are only probabilistic: even if the system is optimal with respect to some definition, the opponent has always a non-zero probability to cheat. However, this probability can be made exponentially small. The advantage of this approach lies in the unconditional security. Like in the case of the Vernam scheme, the price paid for this is that these schemes are rather impractical.

The first result on unconditionally secure authentication appeared in 1974 in a paper by E. Gilbert, F. MacWilliams, and N. Sloane [57]. Subsequently the theory has been developed by G. Simmons, analogous to the theory of secrecy systems that was invented by C. Shannon [137]. An overview of this theory can be found in [138, 139]. From the brief but clear summary by J. Massey in [92] we can cite the following statement *"The theory of authenticity is in many ways more subtle than the corresponding theory of secrecy. In particular, it is not at all obvious how "perfect authenticity" should be defined.* This is caused by the fact that there are different bounds that can be met with equality. In this section we will give the basic definitions, discuss briefly a taxonomy, and give (without proof) the most important theoretical bounds.

Definitions and Notations. According to the definitions that are used in the literature on this subject, the information to be communicated is called the *source state*, while the information that will be sent to the receiver is called the *message*. A mapping between the space of source states and the message space is called an *encoding rule*. The set of source states, messages and encoding rules is called an *authentication code*. In the following we will keep the more usual terminology of plaintext, ciphertext and key. The information theoretic study of authentication has now been reduced to the design of authentication codes, that are in some sense dual to error correcting codes [139]. In both cases redundant information is introduced: in the case of error correcting codes the purpose of this redundancy is to allow the receiver to reconstruct the actual message from the received codeword, and to facilitate this the most likely alterations are in some metric close to the original codeword; in case of authentication codes, the goal of the redundancy is to allow the receiver to detect substitutions or impersonations by an active eavesdropper, and this is obtained by spreading altered or substituted messages as uniformly as possible.

The set of plaintext, ciphertexts and keys will be denoted with $\{P\}$, $\{C\}$, and $\{K\}$ respectively. The size of plaintext, ciphertext, and key space will be denoted with p, c, and k.

In the theory of authenticity, a game with three players is considered like in the theory of secrecy systems. First one has the sender Alice, who wants to send information to the receiver Bob under the form of a cryptogram. The opponent of Alice and Bob is the active eavesdropper Eve, who can perform three types of attacks:

- Eve can send a fraudulent cryptogram to Bob as if it came from Alice (*impersonation attack*).
- Eve can wait until she observes a cryptogram and replace it by a different cryptogram (*substitution attack*).
- Eve can choose freely between both strategies (*deception attack*).

The probability of success (when the strategy of Eve is optimal) will be denoted with P_i, P_s, and P_d respectively. A first result which follows from Kerckhoffs' assumption[2] (namely that the strategy to choose the key is known by Eve) is that

$$P_d = \max(P_i, P_s).$$

Extensions of this model are discussed in [139].

A basic distinction that can be made in this theory is between authentication codes with and without secrecy. In the latter case the plaintext can be derived easily from the ciphertext. The corresponding codes are called *Cartesian*. In this case one can write the ciphertext as the concatenation of the plaintext P with an authenticator, Message Authentication Code, or MAC. The number of authenticators is denoted with r.

Bounds on Authentication Codes. The simplest bound that can be given is for an impersonation attack: if the enemy selects C completely at random from the c cryptograms that occur with nonzero probability in the authentication code, his probability of success can be lower bounded by the following expression:

$$P_i \geq \frac{p}{c},$$

which is called the combinatorial bound for impersonation. One can show that this bound can not be met with equality if splitting occurs. For Cartesian codes this bound reduces to $P_i > 1/r$.

If no splitting occurs, a similar bound can be given for substitution:

$$P_s \geq \frac{p-1}{c-1},$$

[2] The Dutchman A. Kerckhoffs (1853–1903) was the first to enunciate the principle that the security of a cipher must reside entirely in the secret key.

which is called the combinatorial bound for substitution.

The next bound is based on the concept of mutual information, and is also called the result on the authentication channel capacity:

$$P_i \geq 2^{-I(C;K)}.$$

For the shortest proof known until now and a discussion of the recent improvements on this bound by R. Johannesson and A. Sgarro the reader is referred to [92]. A formulation in words is that the difference between the amount of information transmitted through the channel and that needed by the receiver to resolve his uncertainty about the the plaintext can be used to authenticate the plaintext, and conversely no better result can be achieved [139]. One can now show that

$$P_d \geq \frac{1}{\sqrt{k}}. \tag{1}$$

For l-fold secure schemes where an opponent can first observe l ciphertexts, this can be extended to [49]

$$P_d \geq \frac{1}{{}^{l+1}\sqrt{k}}.$$

The next step is to define *perfect authenticity* to mean that equality holds in the equation for the authentication channel capacity. However, even if a system is perfect, P_d will only be small if the cryptogram C reveals much information about the key K. Two simple examples of perfect authentication codes are given in Table 1 [92].

K	$P=0$	$P=1$
00	00	10
01	01	11
10	00	11
11	01	10

K	$P=0$	$P=1$
00	00	10
01	01	00
10	11	01
11	10	11

Table 1. A perfect Cartesian authentication code and a perfect authentication code that also provides perfect secrecy

One can conclude that perfect authenticity schemes are less efficient in terms of key bits than schemes that provide perfect secrecy [91]: if a p'-bit plaintext is mapped to a c'-bit ciphertext, the number of key bits per plaintext bit for perfect authenticity with P_i and P_s equal to the combinatorial bound is at least $(c' - p')/p'$, which is much larger than 1 if p' is small. Note that $P_i = 1/2^{c'-p'}$, so $c' - p'$ can not be small.

Characterization of Perfect Cartesian Authentication Codes. A large number of authentication codes have been proposed in literature that meet these bounds (or a subset of them). We will not attempt to overview all these constructions, but we will concentrate on characterizations. A characterization of an authentication code with certain properties shows that these codes can be obtained essentially in one way, which means that all other constructions are equivalent. For the time being no general characterization of perfect authentication codes is known. For Cartesian codes some characterizations exist in terms of combinatorial structures.

For Cartesian authentication codes that meet the bound (1) with equality, a characterization was given by M. De Soete, K. Vedder, and M. Walker [45] based on nets. This result was in fact a generalization of the work by E. Gilbert, F. MacWilliams, and N. Sloane [57].

For Cartesian authentication codes that meet the combinatorial bound for impersonation and substitution with equality, a characterization was given by D. Stinson [140] based on orthogonal arrays. It should be noted that only for a subset of these codes the bound (1) is met with equality. For general codes that meet the two combinatorial bounds, D. Stinson has derived characterizations based on balanced incomplete block designs.

Practical Cartesian Authentication Codes. In the following more practical authentication schemes will be discussed that are still unconditionally secure. To be more in line with the rest of this paper, the notation will be changed: an m-bit message will be authenticated with a n-bit MAC. The problem with the perfect constructions is that a key of at least $2m$ bits is required. The basic idea to make these schemes more efficient is to allow that P_s increases with an acceptable factor under the assumption that this will reduce the size of the key.

The property of the perfect schemes with minimal key size based on nets [45] is that $P_s = P_i = 2^{-n}$, and the key size is $2n$ bits. One can show that in this case $2^m \leq 2^n + 1$ or $m < n$, and hence the key size is larger than $2m$. In a practical problem, one designs a system with given P_d and size of the plaintext. One can distinguish between three situations:

- For very small messages ($m \ll n$), these schemes are very inefficient because the key size is much larger than twice the message size.
- For medium sized messages ($m \approx n$) the key size will be equal to $2m$, and hence the schemes makes optimal use of the key.
- For large messages ($m \gg n$), it follows that the key size will be equal to about $2m$, but P_d will be much smaller than required. This can be avoided by splitting the message into n bit blocks and authenticating these separately: the efficiency in terms of key bits will remain the same, but the calculations will be on smaller blocks, and hence the scheme will be more efficient [43].

The concept of a universal hash function was introduced by J. Carter and M. Wegman [18, 19]. It can be used to increase the efficiency in terms of the number of key bits at the cost of an increased P_s [22, 141]. Both key size and P_s

will depend on the size of the message and of the MAC. A very elegant scheme was proposed recently by B. den Boer [42]. The key consists of 2 elements of $GF(2^n)$ denoted with μ and ν. The argument x is split into elements of $GF(2^n)$ denoted with $x_1, x_2, \ldots, x_t$, hence $m = t \cdot n$. The function is then defined as follows:

$$g(x) = \mu + \sum_{i=1}^{t} x_i \cdot \nu^i,$$

where the addition and the multiplication are in $GF(2^n)$. It can be shown that this yields an authentication code with $P_i = 2^{-n}$ and $P_s = (m/n)/2^n$. It is clear that this scheme can be generalized to any finite field. Observe that if $m = n$ this construction reduces to the construction by E. Gilbert, F. MacWilliams, and N. Sloane [57].

If these schemes are used in a practical setting, it remains a disadvantage that a single key can be used to authenticate only one message; this can be avoided by encrypting the MAC with the Vernam scheme, which means that n additional key bits per message are required. Other solutions are discussed in [119].

Unconditionally Secure Digital Signatures. Under certain assumptions, it is possible to construct digital signatures that are unconditionally secure for signer or receiver (this implies that even infinite computing power will not enable the other party to cheat with a significant probability). A special case are *fail-stop* signature schemes: the security for the signer is conditional, but if a signature is forged he can prove that the underlying assumption is broken. If the number of participants is limited, one can even construct unconditionally secure signature schemes [21], i.e., digital signatures that are unconditionally secure for both signer and receiver. A detailed discussion of all these schemes is outside the scope of this paper.

3.2 Complexity Theoretic Approach

The approach taken here is to define a model of computation, like a Turing machine [2] or a Boolean circuit. All computations in this model are parameterized by a security parameter, and only algorithms or circuits that require asymptotically polynomial time and space in terms of the size of the input are considered feasible. The next step is then to design cryptographic systems that are provably secure with respect to this model. This research program has been initiated in 1982 by A.C. Yao [145] and tries to base cryptographic primitives on general assumptions. Examples of *cryptographic primitives* are: secure message sending, cryptographically secure pseudo-random generation, digital signatures, and Collision Resistant Hash Functions (CRHF). Examples of *general assumptions* to which these primitives can be reduced are the existence of one-way functions, injections, or permutations, and the existence of trapdoor one-way permutations. A third aspect is the *efficiency of the reduction*, i.e., the number of executions of the basic function to achieve a cryptographic primitive, and the number of interactions between the players in the protocol.

An important research goal is to reduce cryptographic primitives to weaker assumptions, with as final goal to prove that the reduction is optimal. One can also try to improve the efficiency of a reduction, possibly at the cost of a stronger assumption. If someone wants to build a concrete implementation, he will have to choose a particular one-way function, permutation, etc. The properties of a particular problem that is believed to be hard can be used to increase the efficiency of the solutions. Examples of problems that have been intensively used are the factoring of a product of two large primes, the discrete logarithm problem modulo a prime and modulo a composite that is the product of two large primes, and the quadratic residuosity problem.

The complexity theoretic approach has several advantages:

1. It results in *provable secure* systems, based on a number of assumptions.
2. The constructions of such proofs requires *formal definitions* of the cryptographic primitives and of the security of a cryptographic primitive.
3. The *assumptions* on which the security of the systems is based are also defined formally.

The disadvantage is that the complexity theoretic approach has only a limited impact on practical implementations, due to limitations that are inherently present in the models.

1. In complexity theory, a number of operations that is *polynomial* in the size of the input is considered to be feasible, while a superpolynomial or exponential number of operations in the size of the input is infeasible. In an asymptotic setting, abstraction is made from both constant factors and the degrees of the polynomials. This implies that this approach gives no information on the security of concrete instances (a practical problem has a finite size). Secondly, the scheme might be impractical because the number of operations to be carried out is polynomial in the size of the input but impractically large.
2. The complexity theoretic approach yields only results on the *worst case or average case* problems in a general class of problems. However, cryptographers studying the security of a scheme are more interested in the subset of problems that is easy.
3. Complexity usually deals with *single isolated instances* of a problem. A cryptanalyst often has a large collection of statistically related problems to solve.

The most important results on authentication will be summarized briefly. M. Naor and M. Yung have introduced the concept of a Universal One-Way Hash Function (UOWHF) [108]. The philosophy behind a UOWHF is that first the input is selected and subsequently (and independently) the hash function. In this case it does not help an opponent to find collisions for the hash function (cf. Sect. 4.2). They showed how to use a UOWHF to build a signature scheme. An important result is that it is sufficient to have a UOWHF that compresses a single bit to construct a UOWHF that compresses an arbitrary number of bits.

Several authors have subsequently improved their construction. A key result by
J. Rompel [132] is a (very inefficient) construction for a UOWHF based on any
one-way function, which is the weakest possible assumption.

A second important result by I. Damgård [31, 32] is a construction for a
CRHF based on a collision resistant function (with fixed size input).

3.3 System Based or Practical Approach

In this approach schemes with fixed dimensions are designed and studied, paying
special attention to the efficiency of software and hardware implementations.
The goal of this approach is to ensure that breaking a cryptosystem is a difficult
problem for the cryptanalyst.

By trial and error procedures, several *cryptanalytic principles* have emerged,
and the designer intends to avoid attacks based on these principles. Typical
examples are statistical attacks and meet-in-the-middle attacks. An overview of
these principles for cryptographic hash functions will be given in Sect. 8.

The second aspect is to design *building blocks with provable properties*.
These building blocks are not only useful for cryptographic hash functions, but
also for the design of block ciphers and stream ciphers. Typical examples are
statistical criteria, diffusion and confusion, correlation, and non-linearity criteria.

Thirdly, *the assembly of basic building blocks* to design a cryptographic hash
functions can be based on theorems. Results of this type are often formulated and
proven in a complexity theoretic setting, but can easily be adopted for a more
practical definition of "hardness" that is useful in a system based approach. A
typical example is the theorem discovered independently in [33] and [97], stating
that a collision-resistant hash function can always be constructed if a collision-
resistant function exists, where the first reference uses a complexity theoretic
approach and the second a more practical definition.

4 Cryptographic Hash Functions

In Sect. 2, it was explained how the authenticity of information can be verified
through the protection of the secrecy and/or the authenticity of a short imprint
or hashcode. In this section informal definitions will be given for a hash function
that uses a secret key (Message Authentication Code or **MAC**) and for a hash
function that does not make use of a secret key (Manipulation Detection Code
or **MDC**). This last category can be split into two classes, depending on the
requirements: one-way hash functions (OWHF) or weak one-way hash functions
and collision resistant hash functions (CRHF), collision free hash functions or
strong one-way hash functions.

A brief discussion of the existing terminology can avoid confusion that originates from literature. The term *hash functions* originates historically from computer science, where it denotes a function that compresses a string of arbitrary
input to a string of fixed length. Hash functions are used to allocate as uniformly as possible storage for the records of a file. The name hash functions

has also been widely adopted for *cryptographic hash functions* or cryptographically strong compression functions, but the result of the hash function has been given a wide variety of names in cryptographic literature: hashcode, hash total, hash result, imprint, (cryptographic) checksum, compression, compressed encoding, seal, authenticator, authentication tag, fingerprint, test key, condensation, Message Integrity Code (MIC), message digest, etc. The terms MAC and MDC originated from US standards and are certainly not perfect (a MAC or an MDC are actually no codes, and both can serve for message authentication), but the adoption of these terms offers a practical solution to the momentary "Babel of tongues". One example of the confusion is that "checksum" is associated with the well known Cyclic Redundancy Checks (CRC) that are of no use for cryptographic applications. In this paper the names MAC and MDC will also be used for the hashcode obtained with a MAC and an MDC respectively. Sometimes a MAC is called a keyed hash function, but then one has to use for an MDC the artificial term unkeyed or keyless hash function. According to their properties, the class of MDC's will be further divided into one-way hash functions (OWHF) and collision resistant hash functions (CRHF). The term collision resistant hash function (CRHF) is preferable over strong one-way hash function, as it explains more clearly the actual property that is satisfied. The term collision free hash function proposed by I. Damgård is also more explicit, but can be slightly misleading: in fact collisions do exist, but it should be hard to find them. An alternative that was proposed in [148] is collision intractible hash functions. The term weak one-way hash function was proposed by R. Merkle in [97], in order to stress the difference with a strong or collision resistant hash function. Finally note that in a complexity theoretic context the term universal one-way hash function (UOWHF) was proposed by M. Naor and M. Yung in [108] (cf. Sect. 3.2). The relation between the different hash functions has been summarized in Fig. 1.

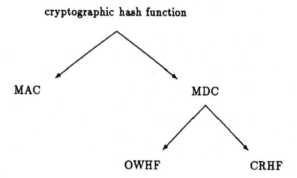

Fig. 1. A taxonomy for cryptographic hash functions

In the following the hash function will be denoted with h, and its argument, i.e., the information to be protected with X. The image of X under the hash function h will be denoted with $h(X)$, and the secret key with K.

4.1 One-Way Hash Function (OWHF)

The first informal definition of a OWHF was apparently given by R. Merkle [95, 97] and M.O. Rabin [125].

Definition 1. *A one-way hash function is a function h satisfying the following conditions:*

1. *The description of h must be publicly known and should not require any secret information for its operation (extension of Kerckhoffs' principle).*
2. *The argument X can be of arbitrary length and the result $h(X)$ has a fixed length of n bits (with $n \geq 64$).*
3. *Given h and X, the computation of $h(X)$ must be "easy".*
4. *The hash function must be one-way in the sense that given X and $h(X)$ it is "hard" to find a message $X' \neq X$ such that $h(X') = h(X)$.*

Note that this last condition (finding a second preimage is "hard") is stronger than the intuitive concept of one-wayness, namely that it is "hard" to find a preimage X given only h and the value of $h(X)$. It is clear that in the case of permutations or injective functions only this concept is relevant. Formal definitions of a OWHF can be obtained through insertion of a formal definition of "hard" and "easy" in combination with the introduction of a security parameter. One has however still several options. In the case of "ideal security", introduced by X. Lai and J. Massey [87], producing a (second) preimage requires 2^n operations. However, it may be that an attack requires a number of operations that is smaller than $O(2^n)$, but is still computationally infeasible.

4.2 Collision Resistant Hash Function (CRHF)

The first formal definition of a CRHF was apparently given by I. Damgård [31, 32]. An informal definition was given by R. Merkle in [97].

Definition 2. *A collision resistant hash function is a function h satisfying the following conditions:*

1. *The description of h must be publicly known and should not require any secret information for its operation (extension of Kerckhoffs' principle).*
2. *The argument X can be of arbitrary length and the result has a fixed length of n bits (with $n \geq 128$).*
3. *Given h and X, the computation of $h(X)$ must be "easy".*
4. *The hash function must be one-way in the sense that given X and $h(X)$ it is "hard" to find a message $X' \neq X$ such that $h(X') = h(X)$.*
5. *The hash function must be collision resistant: this means that it is "hard" to find two distinct messages that hash to the same result.*

Under certain conditions one can argue that the one-way property where the opponent does know the X, follows from the collision resistant property. Similarly formal definitions of a CRHF can be obtained through insertion of a formal definition of "hard" and "easy" in combination with the introduction of a security parameter. For a practical definition, several options are available. In the case of "ideal security" [87], producing a (second) preimage requires 2^n operations and producing a collision requires $O(2^{n/2})$ operations. This can explain why both conditions have been stated separately. One can however also consider the case where producing a (second) preimage and a collision requires at least $O(2^{n/2})$ operations, and finally the case where one or both attacks require less than $O(2^{n/2})$ operations, but the number of operations is still computationally infeasible (e.g., if a larger value of n is selected).

4.3 Message Authentication Code (MAC)

Message Authentication Codes have been used for a long time in the banking community and are thus older than the open research in cryptology that started in the mid seventies. However, MAC's with good cryptographic properties were only introduced after the start of open research in the field.

Definition 3. *A MAC is a function satisfying the following conditions:*

1. *The description of h must be publicly known and the only secret information lies in the key (extension of Kerckhoffs' principle).*
2. *The argument X can be of arbitrary length and the result has a fixed length of n bits.*
3. *Given h, X and K, the computation of $h(K, X)$ must be "easy".*
4. *Given h and X, it is "hard" to determine $h(K, X)$ with a probability of success "significantly higher" than $1/2^n$. Even when a large number of pairs $\{X_i, h(K, X_i)\}$ are known, where the X_i have been selected by the opponent, it is "hard" to determine the key K or to compute $h(K, X')$ for any $X' \neq X_i$. This last attack is called an adaptive chosen text attack.*

Note that this last property implies that the MAC should be both one-way and collision resistant for someone who does not know the secret key K. This definition leaves open the problem whether or not a MAC should be one-way or collision resistant for someone who knows K. An example where this property could be useful is the authentication of multi-destination messages [101].

4.4 Applications of Cryptographic Hash Functions

Cryptographic hash functions can be used to provide for authentication with and without secrecy. It will also be shown that hash functions play a very important role in the design of digital signature schemes. Other applications that will not be discussed here are the authentication of multi-destination messages, the protection of passphrases and the commitment to a string without revealing it.

Authentication Without Secrecy. In this case there is only a plaintext available, which significantly reduces the number of options.

MAC. The simplest approach is certainly the use of a Message Authentication Code or MAC. In order to protect the authenticity of the information, one computes the MAC and appends it to the information. The authenticity of the information now depends on the secrecy and authenticity of the secret key and can be protected and verified by anyone who is privy to this key. Essentially the protection of authenticity has been reduced to the problem of secure key management. This scheme can only protect against outsider attacks, as all parties involved have the same capabilities and hence should trust each other.

The scheme can be made even more secure but less practical if also the authenticity and/or secrecy of the MAC of every plaintext is protected. A possible implementation could consist of an exchange of messages via a high speed communication link, while the corresponding MAC's are sent via a slower channel, that protects authenticity and/or secrecy. A simple authentic channel can be a telephone line (with voice recognition) or the conventional mail system (with manual signatures). The advantage is that it becomes impossible for any of the parties privy to the secret key to modify an existing message and the corresponding MAC.

An issue of discussion is whether it should be "hard" for someone who knows the secret key to construct two arguments with the same MAC for that key. This will strongly depend on the application: in general an internal procedure has to be established to resolve disputes. A third party can make no distinction between the parties involved, but it is possible that, although both parties have the same capabilities, a certain asymmetry exists in their relation, e.g., the bank versus a customer. This procedure should specify what happens if someone denies a messages or subsequently claims he has sent a different message. If some additional protection against insider attacks is obtained from protection of the MAC, this property should be satisfied. However, also physical solutions based on tamper resistant devices can offer such protection. Nevertheless, a better way to solve disputes is to provide for non-repudiation through digital signatures.

MDC. The alternative for a MAC is the use of an MDC. In this case the authenticity of the information is transferred to the authenticity of a string of fixed length. The advantage over a MAC is that there is no need for key management. In exchange for this, an authentic channel has to be provided for every MDC. This means that the capacity of the channel will increase with the number of messages. Although the life time of a key is also related to the number of times it has been used, it is clear that the authentic channel for the MDC will need a significantly greater capacity than the channel that protects both authenticity and privacy of the secret key for a MAC.

Just as in case of a MAC, the parties that use this approach are supposed to trust each other, but it is important to consider what will happen if a dispute arises, or what will happen if an insider will attack the system. An insider will try to find a collision, i.e., two plaintexts X and X' such that $h(X) = h(X')$.

Subsequently he will protect the authenticity of X through $h(X)$, but at any time later he will be able to substitute X' for X. In order to avoid this attack, h should be a CRHF.

However, one can certainly imagine applications where this attack is not relevant. In that case one only has to be protected against outsiders, hence it suffices that h is a OWHF: an outsider can not select X, but will only be able to observe X and $h(X)$ and subsequently try to come up with an X' such that $h(X) = h(X')$.

1. The parties involved completely trust each other, which is trivially the case if there is only one party. One could think of someone who protects the integrity of his computer files through the calculation of an MDC that he stores in printed form in this vault. Every day he can repeat the calculation and verify the result.

2. The computation of the $h(X)$ involves a random component, that can not be controlled by the insider [97]: X can be randomized before applying h through encryption of X with a good block cipher using a truly random key, that is added to the resulting ciphertext [95], or through the selection of a short random prefix to X [31]; h itself can be randomized through randomly choosing h from a family of functions indexed by a certain parameter.

The advantage of a OWHF is that its design is easier and that storage for the hashcode can be halved (64 bits instead of 128 bits). The price paid for this is a degrading of the security level proportional to the number of applications of h: an outsider who knows a set $\{h(X) \mid X \in \text{Domain}(h)\}$ of size s has increased his probability to find an X' with a factor s. This limitation can be overcome through the use of a parameterized OWHF.

Authentication With Secrecy. If both authentication and secrecy are protected, this can be used in certain cases to simplify the overall system. For insider attacks, the additional encryption makes no difference, as an insider knows the secret key for the encryption. This means that for certain applications it should be hard to find collisions. For an outsider, an attack on the scheme becomes in general harder, as his knowledge decreases.

Although it is tempting to use this fact to lower the requirements imposed on the hash functions, this is certainly not a good practice. The additional protection offered by the encryption is dependent on the encryption algorithm and on the mode of the encryption algorithm.

MAC. Several options can be considered, but all share the problem of a double key management: one for the authentication and one for the encryption. It is tempting to use the same key twice, but this has to be discouraged strongly: not only are there dangerous interactions possible between the encryption scheme and the MAC, but the management of both keys should be different (e.g., lifetime, storage after use). The advantage of this approach is a high security level, owing to the complete separation of protection of privacy and authentication.

The most straightforward option is to calculate the MAC, append it to the information and subsequently encrypt the new message. An alternative is to omit the encryption of the MAC. The third solution is to calculate the MAC on the enciphered message. The advantage is that the authenticity can be verified without knowing the plaintext or the secret key of the encryption algorithm, but in general it is preferable to protect the authenticity of the plaintext instead of the authenticity of the ciphertext.

MDC. The advantages for using an MDC are a simplified key management and the fact that the authentication is derived directly from the privacy protection. The key management will be simplified because only one key will be necessary to protect both privacy and authenticity. The fact that the authentication is based on the privacy protection implies that it requires no additional secret key or authentic channel. In the context of the ISO Open System Interconnection Reference Model [71] integrity and confidentiality can be protected at different layers. No secret information would be necessary at the layer that calculates the MDC. The disadvantage is that the protection of authenticity depends on the privacy protection: if the encryption algorithm is weak, the protection of authenticity will also be jeopardized.

The most straightforward option is to calculate the MDC, append it to the information, and subsequently encrypt the new message. An alternative is to omit the encryption of the MDC. This approach seems to be more vulnerable to outsider attacks, but it should cause no problem if the MDC satisfies the imposed conditions. The third solution is to calculate the MDC on the enciphered message. However, this approach can not be recommended: the result has to be protected now with an authentic channel, and an important advantage of this approach is lost.

A special warning should be added in case the encryption algorithm is an additive stream cipher where the opponent knows the plaintext [80, 82]: in that case he can easily compute the key-stream sequence. Subsequently he can modify the plaintext, calculate the MDC and encrypt both plaintext and ciphertext again. This attack depends only on the mode of operation of the encryption and is independent of the choice of the MDC. A solution suggested by R. Jueneman is to let the MDC depend on a random initial value IV that is added to the plaintext, which means that it can not be known by an adversary. This is equivalent to using a MAC and adding the key for the MAC to the plaintext. In a communication environment this could be realized in practice by making the MDC in a message dependent on the previous message and on the previous MDC. The MDC in the first message should be based on a secret random component that was sent by the *receiver* when the session was established. The last message should contain and end-of-session symbol, the MDC of the previous message, and the MDC of the current message. This approach is limited to system where sessions are established in real-time.

5 An Overview of MDC Proposals

5.1 A General Model

Before discussing a small fraction of the large number of proposals for MDC's, the general scheme for describing a hash function will be sketched. Almost all known hash functions are based on a compression function with fixed size input; they process every message block in a similar way. This has been called an "iterated" hash function in [87]. The information is divided into t blocks X_1 through X_t. If the total number of bits is no multiple of the block length, the information can be padded to the required length. The hash function can then be described as follows:

$$H_0 = IV$$
$$H_i = f(X_i, H_{i-1}) \qquad i = 1, 2, \ldots t$$
$$h(X) = H_t.$$

The result of the hash function is denoted with $h(X)$ and IV is the abbreviation for Initial Value. The function f is called the *round* function. Two elements in this definition have an important influence on the security of a hash function: the choice of the padding rule and the choice of the IV. It is recommended that the padding rule is unambiguous (i.e., there exist no two messages that can be padded to the same message), and that it appends at the end the length of the message. The IV should be considered as part of the description of the hash function. In some cases one can deviate from this rule, but this will make the hash function less secure and may lead to trivial collisions or second preimages.

Research on hash functions has been focussed on the question: what conditions should be imposed on f to guarantee that h satisfies certain conditions ? Two main results have been shown on the properties of the round function f. The first result is by X. Lai and J. Massey [87] and gives necessary and sufficient conditions for f in order to obtain an "ideally secure" hash function h.

Proposition 4. *Assume that the padding contains the length of the input string, and that the message X (without padding) contains at least 2 blocks. Then finding a second preimage for h with a fixed IV requires 2^n operations if and only if finding a second preimage for f with arbitrarily chosen H_{i-1} requires 2^n operations.*

A second result by I. Damgård [33] and independently by R. Merkle [97] states that for h to be a CRHF it is sufficient that f is a collision resistant function.

Proposition 5. *Let f be a collision resistant function mapping l to n bits (with $l - n > 1$). If an unambiguous padding rule is used, the following construction will yield a CRHF:*

$$H_1 = f(0^{n+1} \parallel x_1)$$
$$H_i = f(H_{i-1} \parallel 1 \parallel x_i) \qquad \text{for } i = 2, 3, \ldots t.$$

In the following four types of schemes will be briefly discussed: hash functions based on a block cipher, hash functions based on modular arithmetic, hash functions based on a knapsack, and dedicated hash functions. For a more detailed discussion, the reader is referred to [119].

5.2 Hash Functions Based on a Block Cipher

Two arguments can be indicated for designers of hash functions to base their schemes on existing encryption algorithms. The first argument is the minimization of the design and implementation effort: hash functions and block ciphers that are both efficient and secure are hard to design, and many examples to support this view can be found in the literature. Moreover, existing software and hardware implementations can be reused, which will decrease the cost. The major advantage however is that the trust in existing encryption algorithms can be transferred to a hash function. It is impossible to express such an advantage in economical terms, but it certainly has an impact on the selection of a hash function. It is important to note that for the time being significantly more research has been spent on the design of secure encryption algorithms compared to the effort to design hash functions. It is also not obvious at all that the limited number of design principles for encryption algorithms are also valid for hash functions. The main disadvantage of this approach is that dedicated hash functions are likely to be more efficient. One also has to take into account that in some countries export restrictions apply to encryption algorithms but not to hash functions. Finally note that block ciphers may exhibit some weaknesses that are only important if they are used in a hashing mode (cf. Sect. 8.4).

The encryption operation E will be written as $Y = E(K, X)$. Here X denotes the plaintext, Y the ciphertext, and K the key. The size of the plaintext and ciphertext or the block length will be denoted with r, while the key size will be denoted with k. In the case of the well known block cipher DES, $r = 64$ and $k = 56$ [51]. The rate of a hash function based on a block cipher is defined as the number of encryptions to process r plaintext bits.

A distinction will be made between the case $n = r$ and $n = 2 \cdot r$. This is motivated by the fact that most proposed block ciphers have a block length of only 64 bits, and hence an MDC with a result twice the block length is necessary to obtain a CRHF. Other proposals are based on a block cipher with a large key and on a block cipher with a fixed key.

Size of Hashcode Equal to the Block Length. From Definition 2 it follows that in this case the hash function can only be collision resistant if the block length r is at least 128 bits. Many schemes have been proposed in this class, but the first secure schemes were only proposed after several years. Recently the authors have suggested a synthetic approach: we have studied all 64 possible schemes which use exclusive ors and with an internal memory of only one block [119]. As a result, it was shown that 12 secure schemes exist, but up to a linear transformation of the variables, they correspond essentially to 2 schemes : the

1985 scheme by S Matyas, C. Meyer, and J. Oseas [93]:

$$f = E^{\oplus}(s(H_{i-1}), X_i)$$

(here $s()$ is a mapping from the ciphertext space to the key space and $E^{\oplus}(K, X)$ $= E(K, X) \oplus X$), and the variant that was proposed by the authors in 1989 [115] and by Miyaguchi et al. [103] for N-hash and later for any block cipher [78].

$$f = E^{\oplus}(s(H_{i-1}), X_i) \oplus H_{i-1}.$$

The first scheme is currently under consideration for ISO standardization as a one-way hash function [77]. The 12 variants have slightly different properties related to weak keys, the complementation property, and differential attacks [119]. The dual of the first scheme, namely,

$$f = E^{\oplus}(X_i, H_{i-1})$$

is attributed to D. Davies in [144], and to C. Meyer in [37]. Therefore it is denoted with the Davies-Meyer scheme in [102, 123]. It was also shown by the authors that the security level of these hash functions is limited by $\min(k, r)$, even if the size of some internal variables is equal to $\max(k, r)$.

Size of Hashcode Equal to Twice the Block Length. This type of functions has been proposed to construct a collision resistant hash function based on a block cipher with a block length of 64 bits like DES. A series of proposals attempted to double the size of the hashcode by iterating a OWHF; all succumbed to a 'divide and conquer' attack. Another proposal that was broken by the authors [118] is the scheme by Zheng, Matsumoto, and Imai [149]. The Algorithmic Research hash function [79] was broken by I. Damgård and L. Knudsen [34]; a weaker attack was discovered independently by the authors [119].

An interesting proposal was described by R. Merkle [97]. A security "proof" was given under the assumption that DES has sufficient random behavior. However the rate of the most efficient proposal equals about 3.6. The proof for this proposal only showed a security level of 52.5 bits; the authors were able to improve this to 56 bits [119].

Two more efficient schemes called MDC-2 and MDC-4 were proposed by B. Brachtl et al. [14]; these schemes are also known as the Meyer-Schilling hash functions, after the two co-authors who published them at Securicom'88 [100]. For the time being a security proof is lacking. MDC-2 can be described as follows:

$$T1_i = E^{\oplus}(H1_{i-1}, X_i) = LT1_i \parallel RT1_i \quad T2_i = E^{\oplus}(H2_{i-1}, X_i) = LT2_i \parallel RT2_i$$
$$H1_i = LT1_i \parallel RT2_i \qquad\qquad\qquad H2_i = LT2_i \parallel RT1_i$$

Here $H1_0$ and $H2_0$ are initialized with IV_1 and IV_2 respectively, and the hashcode is equal to $H1_t \parallel H2_t$. In order to protect these schemes against attacks based on semi-(weak) keys (cf. Sect. 8.4), the second and third key bits are fixed to 10 and 01 for the first and second encryption. This scheme with rate 2 is currently under consideration for ISO standardization as a CRHF [77]. Finding a collision requires 2^{55} encryptions, and finding a second preimage requires 2^{83} encryptions. One iteration of MDC-4 consists of the concatenation of two MDC-2 steps, where the plaintexts in the second step are equal to $H2_{i-1}$ and $H1_{i-1}$. The rate of MDC-4 is equal to 4. Finding a preimage for MDC-4 requires 2^{109} encryptions, but finding a collision is not harder than in the case of MDC-2. It should be note that the security level of these hash functions might not be sufficient within five to ten years.

Subsequently many attempts were made to improve the efficiency of these proposals. The analysis of all these schemes is rather involved, and depends partially on the properties of the underlying block cipher. Moreover, it can be expected that very efficient schemes are generally more vulnerable. Examples are the scheme suggested in [114], for which the first weakness was identified in [87], and that was finally broken by the authors in [119], and the scheme in [123] that was broken in [27]. Moreover D. Coppersmith has shown that fixed points corresponding to the weak keys of DES are fatal for the schemes in [15, 122]. Both schemes are also vulnerable to an attack based on the complementation property [119]. For the scheme based on LOKI [15], it was already known that it could be broken based on weaknesses of LOKI [9, 40]. These results suggest that countermeasures should be taken to avoid the weaknesses in the block ciphers (e.g., by fixing certain bits), rather than to design hash functions that can deal with these weaknesses.

Size of the Key Equal to Twice the Block Length. Some block ciphers have been proposed for which the key size is approximately twice the block length. Examples in this class are FEAL-NX [104] (a FEAL version with a 128-bit key) and IDEA [88]. Triple DES with 2 keys has a key size of 112 bits and a block length of 64 bits and could hence also be considered to belong to this class.

Size of the Hashcode Equal to the Block Length. A scheme in this class was proposed by R. Merkle in [95]. It can also be classified as "non-invertible chaining":

$$f = E(H_{i-1} \| X_i, IV).$$

An alternative scheme was suggested in [87]:

$$f = E(H_{i-1} \| X_i, H_{i-1}).$$

These constructions can only yield a CRHF if the block length is larger than 128 bits (R. Merkle suggested 100 bits in 1979), and if the key size sufficiently large. For smaller block lengths, a OWHF can be obtained. The security depends strongly on the key scheduling of the cipher.

Size of the Hashcode Equal to Twice the Block Length. In order to obtain a CRHF based on a 64-bit block cipher, a different construction is required. The first two schemes in this class were recently proposed by X. Lai and J. Massey

[87]. Both try to extend the Davies-Meyer scheme. One scheme is called "Tandem Davies-Meyer", and has the following description:

$$T_i = E(H2_{i-1} \| X_i, H1_{i-1})$$
$$H1_i = T_i \oplus H1_{i-1}$$
$$H2_i = E(X_i \| T_i, H2_{i-1}) \oplus H2_{i-1}.$$

The second scheme is called "Abreast Davies-Meyer":

$$H1_i = E(H2_{i-1} \| X_i, H1_{i-1}) \oplus H1_{i-1}$$
$$H2_i = E(H2_{i-1} \| X_i, \overline{H2_{i-1}}) \oplus H2_{i-1}.$$

Both schemes have rate equal to 2, and are claimed to be ideally secure, or finding a pseudo-preimage takes 2^{2n} operations and finding a collision takes 2^n operations.

Schemes With a Fixed Key. All previous schemes modify the key of the block cipher during the iteration phase. The key scheduling process is generally slower than the encryption. Moreover many attacks exploit the fact that the key can be manipulated (e.g., attacks based on weak keys, cf. Sect. 8.4). Finally, this allows to construct a hash function based on any one-way function with small dimensions.

In [117] the authors propose such a scheme with the advantage that a trade-off is possible between security level and speed. The more efficient schemes with a security level of more than 60 bits have a rate equal slightly higher than 4 and need an internal memory of about $3 \cdot 64$ bits. The size of the final hashcode can be reduced by applying a stronger but slower scheme to the final result. The design principles in this paper could be exploited to increase the security level of other hash functions like MDC-2.

5.3 Hash Functions Based on Modular Arithmetic

These hash functions are designed to use the modular arithmetic hardware that is required to produce digital signatures. Their security is partially based on the hardness of certain number theoretic problems. Moreover these schemes are easily scalable. The disadvantage is that precautions have to be taken against attacks that exploit the mathematical structure like fixed points of modular exponentiation (trivial examples are 0 and 1), multiplicative attacks, and attacks with small numbers, for which no modular reduction occurs.

Several schemes with a small modulus (about 32 bits) designed by R. Jueneman (e.g., [81, 82, 83]) have been broken by D. Coppersmith. A second class of schemes uses a large modulus (the size of the modulus n is typically 512 bits or more). In this case the operands are mostly elements of the ring corresponding to an RSA modulus. This poses the following practical problem: the person who has generated the modulus knows its factorization, and hence he has a potential advantage over the other users of the hash function. The solution is that a trusted third party generates the modulus (in that case one can not use the modulus of the user), or one can design the hash function in such a way that the advantage is limited.

The most efficient schemes are based on modular squaring. Moreover some theoretical results suggest that inverting a modular squaring without knowledge of the factorization of the modulus is a difficult problem. Again one can study all possible schemes which use a single squaring and exclusive ors and with an internal memory of only one block. Several schemes of this type have been evaluated in previous papers [59, 112]. The same approach as in Sect. 5.2 can be applied [119]. It shows that the optimal scheme is of the form: $f = (X_i \oplus H_{i-1})^2 \bmod N \oplus H_i$. Most existing proposals use however the well known CBC mode. In order to avoid the vulnerabilities (one can go backwards easily), additional redundancy is added to the message. The first proposal was to fix the 64 most significant bits to 0 [37]. It was however shown in [59, 84] that this is not secure. In a new proposal, that appeared in several standards (e.g., the informative annex D of CCITT-X.509 [20]) the redundancy was dispersed. D. Coppersmith showed however that one can construct two messages such that their hashcode is a multiple of each other [26]. If the hash function is combined with a multiplicative signature scheme like RSA [128], one can exploit this attack to forge signatures [26]. As a consequence, new methods for adding redundancy were proposed within ISO and in [85], but these methods are still under study. Another possibility is the use of the redundancy scheme of [74].

B. den Boer [39] has found collisions for the round function of the squaring scheme by I. Damgård [33], and it was shown in [118] that the scheme by Y. Zheng, T. Matsumoto, and H. Imai [149] is vulnerable to the attack described in [59].

Stronger schemes have been proposed that require more operations. Examples are the use of two squaring operations [59]: $f = \left(H_{i-1} \oplus (X_i)^2 \right)^2 \bmod N$, and the replacement of the squaring by a higher exponent (3 or $2^{16} + 1$) in the previous schemes. This allows to simplify the redundancy [59].

One can conclude that it would be desirable to find a secure redundancy scheme for a hash function based on modular squaring, and to replace the CBC mode by a more secure mode. If a slower scheme is acceptable, the exponent can be increased.

This section will be concluded with provably secure schemes. I. Damgård [31] has suggested constructions for which finding a collision is provably equivalent to factoring an RSA modulus or finding a discrete logarithm modulo a large prime. The construction of J.K. Gibson [58] yields a collision resistant function based on the discrete logarithm modulo a composite. Both the factoring and the discrete logarithm problem are believed to be difficult number theoretic problems. The disadvantages of these schemes is that they are rather inefficient.

5.4 Hash Functions Based on a Knapsack

The knapsack problem was used in 1978 by R. Merkle and M. Hellman to construct the first public key system [94]. However almost all public key schemes based on the knapsack problem have been broken [44], which has given the knapsack a bad reputation. It is an open problem whether the knapsack problem is only hard in worst case, while the average instance is easy. If this would be true, the knapsack problem would be useless for cryptography. The attractivity of the problem lies in the fact that both hardware and software implementations are very fast compared to schemes based on number theoretic problems.

In the case of additive knapsacks, several constructions have been suggested and broken (e.g., P. Camion and J. Patarin have demonstrated in [17] that a second preimage can be constructed for the round function of the scheme by I. Damgård [33]). Other results can be found in [61, 70]. It is for the time being an open problem whether a random knapsack with $n = 1024$ and $b = 512$ is hard to solve. Another solution is to combine the knapsack with a simple additional operation. Also for this hash function one has the problem of trapdoors that can be inserted when choosing the knapsack.

The first multiplicative knapsack proposed by J. Bosset [13] was broken by P. Camion [16]. A new scheme by G. Zémor is also based on the hardness of finding short factorizations in certain groups [147]. For the suggested parameters it can be shown that two messages with the same hashcode will differ in at least 215 bits. It remains an open problem whether it is easy to find factorizations of a 'reasonable size'.

5.5 Dedicated Hash Functions

In this section some dedicated hash functions will be discussed, i.e., algorithms that were especially designed for hashing operations.

MD2 [86] is a hash function that was published by R. Rivest of RSA Data Security Inc. in 1990. The algorithm is software oriented yet not very fast in software. Reduced versions of MD2 (i.e., with less rounds) were shown to be vulnerable [119].

A faster algorithm by the same designer is MD4 [129, 130]. Attacks on reduced versions of MD4 have been developed by R. Merkle and by B. den Boer and A. Bosselaers [39]. This resulted in a strengthened version of MD4, namely MD5 [131]. It was however shown by B. den Boer and A. Bosselaers [41] that the round function of MD5 is not collision resistant. This does not yield a direct attack, but it raises some doubts about the security: one of the design goals, namely design a collision resistant function is not satisfied. A second improved variant of MD4, the Secure Hash Algorithm, was proposed by NIST [55]. The size of the hashcode is increased from 128 to 160 bits and the message words are not simply permuted but encoded with a cyclic code. Another improved version of MD4 called RIPEMD was developed in the framework of the EEC-RACE project RIPE [127]. HAVAL was proposed by Y. Zheng, J. Pieprzyk, and J. Seberry at Auscrypt'92 [150]; it is an extension of MD5.

N-hash is a hash function designed by S. Miyaguchi, M. Iwata, and K. Ohta [103, 105]. Serious weaknesses have been identified in this scheme by B. den Boer [40] and E. Biham and A. Shamir [8]. A new version of N-hash appeared in a Japanese contribution to ISO [78]. It was shown that this scheme contains additional weaknesses [11, 119].

FFT-Hash I and II are MDC's suggested by C.P. Schnorr [135, 136]. The first version was broken independently by J. Daemen, A. Bosselaers, R. Govaerts, and J. Vandewalle [29] and by T. Baritaud, H. Gilbert, and M. Girault [6]. The second version was broken three weeks after its publication by S. Vaudenay [142].

R. Merkle suggested in 1989 a software oriented one-way hash function called Snefru [98]. It is based on large random substitution tables (2 Kbyte per pass). E. Biham and A. Shamir have shown that Snefru [9] is vulnerable to differential attacks. As a consequence it is recommended to use 6 and preferably 8 passes, possibly combined with an increased size of the hashcode. However, these measures increase the size of the substitution tables and decrease the performance.

The scheme by I. Damgård [33] based on a cellular automaton was broken by J. Daemen, J. Vandewalle, and R. Govaerts in [28]. In the same paper these authors have proposed Cellhash, a new hash function based on a cellular automaton [28]. Later an improved version called Subhash was published [30]. Both schemes are hardware oriented.

6 An Overview of MAC Proposals

The general model for an iterated MAC is similar as the model for an MDC. The basic difference is that the round function f and in some cases the initial value IV depend on the secret key K.

In contrast with the variety of MDC proposals, very few algorithms exist. This can perhaps be explained by the fact that the existing standards are still widely accepted. The ANSI standard [4] specifies that the resulting MAC contains 32 bits. It is clear that a result of 32 bits can be sufficient if a birthday attack (cf. Sect. 8) is not feasible and if additional protection is present against random attacks (cf. Sect. 8), which is certainly the case in the wholesale banking environment. In other applications, this can not be guaranteed. Therefore, certain authors recommend also for a MAC a result of 128 bits [82, 83].

The most widespread method to compute a MAC are the Cipher Block Chaining (CBC) and Cipher FeedBack (CFB) mode of DES [4, 53, 73, 75, 99]. The descriptions and standards differ because some of them select one of the two modes, suggest other padding schemes or leave open the number of output bits that is used for the MAC.

$$CBC : f = E(K, H_{i-1} \oplus X_i) \quad \text{and} \quad CFB : f = E(K, H_{i-1}) \oplus X_i .$$

In the case of CFB it is important to encrypt the final result once more, to avoid a linear dependence of the MAC on the last plaintext block.

In the case of DES, an attack based on exhaustive key search (cf. Sect. 8) or differential attacks [10] can be thwarted by encrypting only the last block with triple DES; at the same time this can block the following chosen plaintext attack [40] (it will be described for the case of CBC): let H and H' be the CBC-MAC corresponding to key K and plaintext X and X' respectively. The attacker appends a block Y to X and obtains with a chosen plaintext attack the new MAC, that will be denoted with G. It is then clear that the MAC for the concatenation of X' and $Y' = Y \oplus H \oplus H'$ will also be equal to G. An alternative way to block this attack is to encrypt the result with a key derived from K.

If both authenticity and secrecy are protected using the same block cipher, the keys for both operations have to be different [80, 99, 119].

A new mode to compute a MAC was suggested by the authors [119]:

$$f = E(K, X_i \oplus H_{i-1}) \oplus X_i \, .$$

It has the advantage that the round function is harder to invert.

The Message Authentication Algorithm (MAA) is a dedicated MAC. It was published in 1983 by D. Davies and D. Clayden in response to a request of the UK Bankers Automated Clearing Services (BACS) [36, 38]. In 1987 it became a part of the ISO 8731 banking standard [73]. The algorithm is software oriented and has a 32-bit result, which makes it unsuitable for certain applications.

A new non-iterative MAC based on stream ciphers was proposed recently by X. Lai, R. Rueppel and J. Woollven [89]. Further study is necessary to assess its security.

The DSA algorithm (Decimal Shift and Add, not to be confused with the Digital Signature Algorithm) was designed in 1980 by Sievi of the German Zentralstelle für das Chiffrierwesen, and it is used as a message authenticator for banking applications in Germany [38]. Weaknesses of this algorithm have been identified in [67, 119]. The scheme by F. Cohen [23] and its improvement by Y. Huang and F. Cohen [69] proved susceptible to an adaptive chosen message attack [116]. Attacks were also developed [119] on the weaker versions of this algorithm that are implemented in the ASP integrity toolkit [24]. Several MAC algorithms exist that have not been published, such as the S.W.I.F.T. authenticator and Dataseal [90].

Finally it should be noted that if one is willing to exchange a very long key, one should consider the unconditionally secure schemes suggested in Sect. 3.1.

7 Performance of Hash Functions

In order to compare the performance of software implementations of hash functions, an overview has been compiled in Table 2. All timings were performed on a 16 MHz IBM PS/2 Model 80 with a 80386 processor. The implementations were written by A. Bosselaers. Most of them use additional memory to improve the speed. The C-code was compiled with a 32-bit compiler in protected mode. The table has been completed with the speed of DES, a modular squaring and a modular exponentiation. For the last two operations a 512-bit modulus was chosen, and no use was made of the Chinese remainder theorem to speed up the computations. From these figures it can be derived that MDC-2 will run at about 100 Kbit/sec. Some algorithms like Snefru and SHA would perform relatively better on a RISC processor, where the complete internal state can be stored in the registers. On this type of processor, SHA is only about 15% slower than MD5.

type	hash function	C language (Kbit/sec)	Assembly language (Kbit/sec)
MAC	MAA		2750
MDC	MD2	78	78
	MD4	2669	6273
	MD5	1849	4401
	SHA	710	1370
	RIPEMD	1334	3104
	N-hash	266	477
	FFT-hash I	212	304
	Snefru-6	358	358
	Snefru-8	270	270
block cipher	DES (+ key schedule)	130	200
	DES (fixed key)	512	660
modular	squaring	50	273
arithmetic	exponentiation $(2^{16}+1)$	1.8	14

Table 2. Performance of several hash functions on an IBM PS/2 (16 MHz 80386)

8 Methods of Attack

The goal of this section is to give an overview of the known methods of attack on hash functions. This taxonomy can be helpful to understand the attacks published in the literature, but can also serve as a 'caveat' for designers or evaluators of hash functions. The evaluation of an authentication protocol or a signature scheme strongly depends on the information at the disposal of an adversary, the actions he can undertake, and finally on the consequences of both a successful and an unsuccessful attack.

In general, a conservative approach is recommended. This implies that for a MAC and for a digital signature one requires that it should not be existentially forgeable under an adaptive chosen message attack. This means that an attacker can choose a large number of messages and obtain the corresponding MAC (or signature); 'adaptive' means that the choice of the ith message may depend on the outcome of the previous experiments. Finally he has to come up with a message (different from the messages in the attack) and the corresponding MAC (or signature); 'existential' means that the attacker has no control over this message, that might be random or nonsensical.

8.1 Attacks Independent of the Algorithm

These attacks are applicable to any hash function and influence the choice of the parameters (size of hashcode and possibly size of key). For the time being 2^{56} operations is considered to be on the edge of feasibility. In view of the fact that the speed of computers is multiplied by four every three years, 2^{64} operations is sufficient for the next 10 years, but it will be only marginally secure within 20 years. For applications with a time frame of 20 years or more, one should try to design the scheme such that an attack requires at least 2^{80} operations.

Random Attack. The opponent selects a random message and hopes that the change will remain undetected. In case of a good hash function, his probability of success equals $1/2^n$ with n the number of bits of the hashcode. The feasibility of this attack depends on the action taken in case of detection of an erroneous result, on the expected value of a successful attack, and on the number of attacks that can be carried out. For most application this implies that $n = 32$ bits is not sufficient.

Birthday Attack. The idea behind the birthday attack [146] is that for a group of 23 people the probability that at least two people have a common birthday exceeds $1/2$. Intuitively one would expect that the group should be significantly larger. This can be exploited to attack a hash function in the following way: an adversary generates r_1 variations on a bogus message and r_2 variations on a genuine message. The probability of finding a bogus message and a genuine message that hash to the same result is given by $1 - \exp(-\frac{r_1 \cdot r_2}{2^n})$, which is about 63 % when $r = r_1 = r_2 = 2^{\frac{n}{2}}$. Note that in case of a MAC the opponent is unable to generate the MAC of a message. He could however obtain these MAC's with a chosen plaintext attack. A second possibility is that he collects a large number of messages and corresponding MAC's and divides them into two categories, which corresponds to a known plaintext attack. The involved comparison problem does not require r^2 operations: after sorting the data, which requires $O(r \log r)$ operations, comparison is easy. Jueneman has shown [82] that for $n = 64$ the processing and storage requirements are feasible in reasonable time with the computer power available in every large organisation. A time-memory-processor trade-off is possible.

If the function can be called as a black box, one can use the collision search algorithm proposed by J.-J. Quisquater [121], that requires about $2\sqrt{\pi/2} \cdot 2^{\frac{n}{2}}$ operations and negligible storage. To avoid this attack with a reasonable safety margin, n should be at least 128 bits. This explains the second condition in Definition 2 of a CRHF.

In case of digital signatures, a sender can attack his own signature or the receiver or a third party could offer the signer a message he's willing to sign and replace it later with the bogus message. Only the last attack can be thwarted through randomizing the message just prior to signing. If the sender attacks his own signature, the occurrence of two messages that hash to the same value might make the signer suspect, but it will be very difficult to prove to a third party the denial.

Exhaustive Key Search. This attack is only relevant in case of a MAC. It is a known plaintext attack, where an attacker knows M plaintext-MAC pairs for a given key and will try to determine the key by trying all possible keys. In order to determine the key uniquely, M has to be slightly larger than k/n. The expected number of trials is equal 2^{k-1}, with k the size of the key in bits.

8.2 Attacks Dependent on the Chaining

This class of attacks depends on some high level properties of the round function f.

Meet in the Middle Attack. This attack is a variation on the birthday attack, but instead of the hashcode, intermediate chaining variables are compared. The attack enables an opponent to construct a message with a prespecified hashcode, which is not possible in case of a simple birthday attack. Hence it also applies to a OWHF. The opponent generates r_1 variations on the first part of a bogus message and r_2 variations on the last part. Starting from the initial value and going backwards from the hashcode, the probability for a matching intermediate variable is given by the same formula. The only restriction that applies to the meeting point is that it can not be the first or last value of the chaining variable. The cycle finding algorithm by J.-J. Quisquater can be extended to perform a meet in the middle attack with negligible storage [124]. The attack can be thwarted by avoiding functions f that are invertible to the chaining variable H_{i-1} and to the message X_i.

Further extensions of this attack have been proposed in [25, 60] to break p-fold iterated schemes, i.e., weak schemes with more than one 'pass' over the message as proposed in [35]. Other extensions take into account additional constraints on the message.

Correcting Block Attack. This attack consists of substituting all blocks of the message except for some block X_j. This block is then calculated such that the hashcode takes a certain value, which makes it also suitable to attack a OWHF. It often applies to the last block and is then called a correcting last block attack, but it can also apply to the first block or to some blocks in the middle. The hash functions based on modular arithmetic are especially sensitive to this attack.

A correcting block attack can also be used to produce a collision. One starts with two arbitrary messages X and X' and appends one or more correcting blocks denoted with Y and Y', such that the extended messages $X\|Y$ and $X'\|Y'$ have the same hashcode.

One can try to thwart a correcting block attack by adding redundancy to the message blocks, in such a way that it becomes computationally infeasible to find a correcting block with the necessary redundancy. The price paid for this solution is a degradation of the performance.

Fixed Point Attack. The idea of this attack is to look for a H_{i-1} and X_i such that $f(X_i, H_{i-1}) = H_{i-1}$. If the chaining variable is equal to H_{i-1}, it is possible to insert an arbitrary number of blocks equal to X_i without modifying the hashcode. Producing collisions or a second preimage with this attack is only possible if the chaining variable can be made equal to H_{i-1}: this is the case if IV can be chosen equal to a specific value, or if a large number of fixed points can be constructed (if e.g., one can find an X_i for every H_{i-1}). Of course this attack can be extended to fixed points that occur after a number of steps. This attack can be prevented easily: one can append a block count to the data or one can (for theoretical constructions) encode the data with a prefix-free code [33].

Key Collisions. A recent breakthrough in the analysis of block cipher algorithms and hash functions based on block ciphers was a simple collision search algorithm by J.-J. Quisquater [121]. A collision is a pair of keys K_1, K_2 such that $E(K_1, P) = E(K_2, P)$ for a plaintext P. In the case of DES, there exist about 2^{48} collisions for a given P. The running time of the algorithm is $O(2^{\frac{n}{2}})$ and it requires very little storage. The attack can be extended to the case of double encryption [123]. The collision search is applicable to any block cipher with $n < 128$, but a good design of the hash function can make the collisions useless. There is however no easy way to guarantee this, and every scheme has to be verified for this attack.

Differential Attacks. Differential cryptanalysis is based on the study of the relation between input and output differences and is applicable to both block ciphers and hash functions [7, 8]. Interesting results were obtained for FEAL, DES, N-hash and Snefru. Differential attacks of hash functions based on block ciphers have been studied in [119].

Analytical Weaknesses. Some schemes allow manipulations as insertion, deletion, permutation, and substitutions of blocks. A large number of attacks have been based on a blocking of the diffusion of the data input: this means that changes have no effect or can be cancelled out easily in a next stage. This type of attacks has been successful for dedicated hash functions [6, 9, 28, 29, 142] and for hash functions based on modular arithmetic [113].

8.3 Attacks Dependent on Interaction with the Signature Scheme

In some cases it is possible that even if the hash function is collision resistant, it is possible to break the signature scheme. This attack is then the consequence of a dangerous interaction between both schemes. In the known examples of such an interaction both the hash function and the signature scheme have some multiplicative structure [26].

It was shown in [31] that the security of a digital signature scheme which is not existentially forgeable under a chosen message attack will not decrease if it is combined with a CRHF.

8.4 Attacks Dependent on the Underlying Block Cipher

Certain weaknesses of a block cipher are not significant when it is used to protect the privacy, but can have dramatic consequences if the cipher is used in one of the special modes for hashing. These weaknesses can be exploited to insert special messages or to carry out well chosen manipulations without changing the hash result. We will only discuss the weaknesses of DES [3, 51], because of its widespread use.

Complementation Property. One of the first properties that was known of DES was the symmetry under complementation [68]:

$$\forall P, K : C = \mathrm{DES}(K, P) \Longleftrightarrow \overline{C} = \mathrm{DES}(\overline{K}, \overline{P})$$

It can reduce an exhaustive key search by a factor 2 but it also allows to construct trivial collisions.

Weak Keys. Another well known property of DES is the existence of 4 weak keys [38, 106]. For these keys, encryption equals decryption, or DES is an involution. These keys are also called palindromic keys. This means that $E(K, E(K, P))$ $= P, \forall P$. There exist also 6 pairs of semi-weak keys, for which $E(K_2, E(K_1, P))$ $= P, \forall P$. This property can be exploited in certain hash functions to construct fixed points after two iterations steps.

Fixed Points. Fixed points of a block cipher are plaintexts that are mapped to themselves for a certain key. As a secure block cipher is a random permutation, it will probably have fixed points (for every key there is a probability of $1 - e^{-1}$ that there is at least a single fixed point). However, it should be hard to find these. In the case of DES, finding fixed points is easy for some keys [106]: for every weak key K_p, there exist 2^{32} values of P that can be easily found for which $\mathrm{DES}(K_p, P) = P$. A similar property holds for the anti-palindromic keys: these are 4 semi-weak keys for which there exist 2^{32} values of P that can be easily found for which $\mathrm{DES}(K_{ap}, P) = \overline{P}$.

8.5 High Level Attacks

Even if the above attacks would not be feasible, special care has to be taken to avoid replay of messages and construction of valid messages by cutting and splicing others. Also the timeliness and the order of messages can be important.

Attacks on this level can be thwarted by adding time stamps, serial numbers or random challenges and through the use of sound cryptographic protocols. In the case of stored information, a restore can be avoided through the use of version numbers and the order of the blocks can be protected through adding the memory address to the information before the computation of the redundancy.

9 Digital Signature Schemes

In this section some practical constructions for digital signatures will be discussed. The most popular constructions are based on public-key techniques, but one can also construct digital signatures based on physical assumptions and based on conventional one-way functions.

9.1 Based on Physical Assumptions

A device like a smart card is called tamper resistant if one believes that it is hard to access the secret key stored in it. One can use a smart card with a conventional cryptographic algorithm to build a signature scheme as follows: the 'signer' has a smart card that can only encipher with a secret key K, and each 'verifier' has a smart card that can only decipher with a secret key K. Forging a signature is hard if one believes that the devices are tamper resistant. The disadvantage of this approach is that the secret key K has to be installed and stored securely within the smart card of both signer and verifier(s). This should be contrasted with the schemes of Sect. 9.3 where only the secret key of the signer is protected in a smart card.

9.2 Based on a Conventional One-Way Function

The first digital signature scheme based on a conventional one-way function (denoted with g) is the Diffie-Lamport one time signature [46]. In order to sign a single bit message, the sender randomly selects a pair $x_1, x_2 \in \text{Dom}(g)$ and computes $y_1 = g(x_1)$ and $y_2 = g(x_2)$. Next he puts y_1, y_2 in an authenticated public file. If he wants to sign the message, he reveals x_1 if the message bit equals 0 and x_2 if the message bit equals 1. Subsequently the receiver can verify that $y_i = g(x_i)$. A second well known construction is the Rabin scheme [125]; the 'cut and choose' technique proves to be a very powerful building block for other cryptographic protocols.

The main disadvantage of these schemes are the size of keys and signatures and the fact that they can only be used for a fixed number of signatures (often only once). The optimizations by R. Merkle [96] yield a scheme that is almost practical. These basic schemes have served as a starting point for complexity theoretic constructions.

9.3 Based on Public-Key Techniques

An elegant solution to construct digital signatures was proposed by W. Diffie and M. Hellman in their seminal 1976 paper [46] in which they introduced the concept of trapdoor one-way functions. These functions are easy to compute in one direction and difficult to compute in the other direction, except for someone who knows the 'trapdoor' information. Information can then be digitally *signed* if the sender transforms the information with his secret key (the trapdoor information). The receiver can then *verify* the digital signature by applying the transformation in the 'easy direction' using the public key. The signature is unique to the sender because he is the only one that has the secret information.

The same trapdoor one-way functions can also be used for public-key encryption systems, where the receiver can make his key public through an integrity protected channel. The encryption operation is then the transformation with the public key, and the decryption the inverse transformation (using the secret key or trapdoor).

The first 'secure' trapdoor one-way permutation was proposed by R. Rivest, A. Shamir, and L. Adleman in 1977 [128]. The RSA-public key cryptosystem is based on modular exponentiation and its security relies on the difficulty of factoring integers that are the product of two large primes. It is widely used and has become a "de facto" public-key standard. A variant of RSA that is provably equivalent to factoring was proposed by M.O. Rabin [126]; it is based on modular squaring. The RSA and the Rabin scheme are the only public-key signature schemes that have a deterministic signing procedure, i.e., the signing process does not need any random bits. Other efficient variants have been proposed based on the fact that even finding a good approximation for a modular square root is hard if the factorization of the modulus is not known. An example in this class is ESIGN [56]; this scheme is based on a modulus of the form $n = p^2 q$.

A second class of schemes is based on the discrete logarithm problem. The first scheme in this class was suggested by T. ElGamal in 1984 [47]. An optimized variant was put forward in [1]. The draft standard Digital Signature Algorithm proposed by NIST [54] also belongs to this class. It is based on the hardness of the discrete logarithm in $GF(p)$, where p is a large prime number (512 to 1024 bits) such that $p-1$ has a 160-bit prime divisor q. The scheme exploits this structure to optimize storage and computation as suggested by C.P. Schnorr [134] for a different scheme. The advantage of schemes based on the discrete logarithm is that they can be implemented in any group for which the discrete logarithm is believed to be hard. Recently some promising results have been achieved with schemes based on elliptic curves over finite fields (e.g., [66]).

A third class of schemes is derived from zero-knowledge identification protocols; the schemes can again be divided into two classes according to the underlying assumption. The variants based on factoring are Guillou-Quisquater [120] and Ohta-Okamoto [109], and a variant with exponent 2 is the Feige-Fiat-Shamir scheme [50]. The variants based on the discrete logarithm problem are the schemes by Schnorr [134] and Okamoto [110].

The schemes that have been mentioned in this section are all reasonably efficient, but they do not have the same level of provable security, i.e., the relation with the underlying hard problem might be different. The schemes for which an even stronger reduction is possible (e.g., [12, 62]) or the schemes based on more general assumptions (e.g., [108, 132]) are however less practical.

9.4 Combining a Hash Function and a Signature Scheme

The basic idea to speed up all digital signature schemes is to compress the information to be signed with an MDC to a string of fixed length. The procedure will be described for a digital signature scheme with recovery like RSA. The procedures for other signature schemes are similar. Instead of signing the information one signs the hashcode, that is called the "imprint" in standards terminology. In order to verify a signature, the outcome of the verification process of the signature is compared to the hashcode that can be calculated from the information. The advantages of this approach are the following:

1. The size of the signature can be reduced from the size of the information to one block length, independent of the length of the signed information.
2. The sign and verify function of most known signature schemes are several orders of magnitude slower in hardware and software than symmetric encryption functions, MAC's or MDC's (cf. Sect. 7).
3. If information is signed that is longer than one block, it is easy to manipulate these individual blocks. The simplest example is a reordering of blocks.
4. The algebraic structure of the message space can be destroyed. In the case of RSA [48] and discrete logarithm based signatures, the message space has a multiplicative structure, i.e., the signature of the product of two messages equals the product of their signatures. Examples of how this algebraic structure can be exploited in a protocol are described in [63, 107].
5. The reblocking problem can be avoided. This problem occurs when both privacy and authentication are protected with the same RSA-like public-key cryptosystem. If the sender first encrypts the message with the public key of the receiver (to protect privacy), the result might be larger than his modulus. Before he can apply his secret key (to protect authenticity) he has to reblock the message. It is not too hard to overcome the reblocking problem: a simple solution is to let the order of the operations depend on the size of the two moduli. However, it is preferable to have no reblocking problem at all.
6. The signature protocol will not be useful for an opponent trying to obtain the plaintext corresponding to encrypted messages. This can only happen if one uses the same public key cryptosystem and modulus for privacy protection and authentication.

For the zero-knowledge based signature schemes, the hash function is an integral part of the signing and verification operations, as it links together the information to be signed and some unpredictable and signature dependent information. This implies that it is sufficient that the hash function is a OWHF, while for other schemes one needs a CRHF.

In some cases their can be an unfortunate interaction between the digital signature scheme and the hash function (cf. Sect. 8.3). To thwart this problem, one can impose additional conditions on the hash function (e.g., it should be non-homomorphic, correlation free [110],...). An alternative approach, which seems preferable to the authors, is to solve the problems with the signature scheme: an example is the redundancy scheme for RSA (and its extension to even exponents) which was standardized in [74]: it destroys the homomorphic property as explained in [65].

9.5 Selecting a Signature Scheme

When selecting a particular signature scheme, one has to consider the following aspects:

efficiency in terms of storage and computation: number of operations for signing and verifying, size of secret and public key, size of the signature, and the possibility of preprocessing. An extensive comparison can be found in [111].

random: randomized signatures will probably leak less information about the secret key, but on the other hand the signer needs to produce random bits in a secure way (if the random input is revealed, not only the signature but the secret key might be compromised).

security: is the scheme well established, and how strong is the reduction to the underlying hard problem?

hash function: the requirements to be imposed on the hash function.

patents and standards: these 'political' aspects will also influence the choice of a particular scheme.

9.6 Some Practical Considerations

Digital signature schemes have been criticized seriously in the past; it is certainly true that some problems exist, but this is no reason to throw out the baby with the bathwater. An important problem is a user who looses his secret key, or who claims he has lost his secret key. In both cases he will try to revoke his digital signatures. Any practical signature scheme has to implement procedures for the revocation of keys. In order to minimize this risk, one can store the secret key in a tamper resistant device (e.g., a smart card), such that even the owner does not know it, and one can maintain black lists for users with a bad history. This will make fraudulent revocations very difficult. On the other hand, for very large amounts one has to use arbitrated signature schemes, where a third party is involved in the signing process. Another major problem is the legal validity of digital signatures [5]. Manual signatures are certainly not perfect, but it remains an open problem whether at present a judge will accept a digital signature. Legal support of digital signatures is certainly necessary to get a wide acceptance and will be a big step forward for the automation of business transactions.

10 Conclusions

For protecting information authentication, several cryptographic techniques are available: MAC, MDC, digital signature. For a MAC, efficient standards exist that are widely accepted. The design of MDC's for symmetric integrity protection and for digital signatures is apparently more complicated, and new proposals that are both efficient and secure are still under study. The concept of digital signatures is very promising. At present, first standards on the subject are emerging and commercial applications become feasible. Further theoretical developments together with legal support will make digital signatures widely accepted and used in the nineties.

References

1. G.B. Agnew, R.C. Mullin, and S.A. Vanstone, "Common application protocols and their security characteristics," *CALMOS CA34C168 Application Notes*, U.S. Patent Number 4,745,568, August 1989.

2. A.V. Aho, J.E. Hopcroft, and J.D. Ullman, *"The Design and Analysis of Computer Algorithms,"* Addison-Wesley, 1974.

3. ANSI X3.92-1981, *"American National Standard for Data Encryption Algorithm (DEA),"* ANSI, New York.

4. ANSI X9.9-1986 (Revised), *"American National Standard for Financial Institution Message Authentication (Wholesale),"* ANSI, New York.

5. M. Antoine, J.-F. Brakeland, M. Eloy, and Y. Poullet, "Legal requirements facing new signature technology," *Advances in Cryptology, Proc. Eurocrypt'89, LNCS 434*, J.-J. Quisquater and J. Vandewalle, Eds., Springer-Verlag, 1990, pp. 273-287.

6. T. Baritaud, H. Gilbert, and M. Girault, "FFT hashing is not collision-free," *Advances in Cryptology, Proc. Eurocrypt'92, LNCS 658*, R.A. Rueppel, Ed., Springer-Verlag, 1993, pp. 35-44.

7. E. Biham and A. Shamir, "Differential cryptanalysis of DES-like cryptosystems," *Journal of Cryptology*, Vol. 4, No. 1, 1991, pp. 3-72.

8. E. Biham and A. Shamir, "Differential cryptanalysis of Feal and N-hash," *Advances in Cryptology, Proc. Eurocrypt'91, LNCS 547*, D.W. Davies, Ed., Springer-Verlag, 1991, pp. 1-16.

9. E. Biham and A. Shamir, "Differential cryptanalysis of Snefru, Khafre, REDOC-II, LOKI, and Lucifer," *Advances in Cryptology, Proc. Crypto'91, LNCS 576*, J. Feigenbaum, Ed., Springer-Verlag, 1992, pp. 156-171.

10. E. Biham and A. Shamir, "Differential cryptanalysis of the full 16-round DES," *Technion Technical Report # 708*, December 1991.

11. E. Biham, "On the applicability of differential cryptanalysis to hash functions," *E.I.S.S. Workshop on Cryptographic Hash Functions*, Oberwolfach (D), March 25-27, 1992.

12. J. Bos and D. Chaum, "Provably unforgeable signatures," *Advances in Cryptology, Proc. Crypto'92, LNCS*, E.F. Brickell, Ed., Springer-Verlag, to appear.

13. J. Bosset, "Contre les risques d'altération, un système de certification des informations," *01 Informatique*, No. 107, February 1977.

14. B.O. Brachtl, D. Coppersmith, M.M. Hyden, S.M. Matyas, C.H. Meyer, J. Oseas, S. Pilpel, and M. Schilling, *"Data Authentication Using Modification Detection Codes Based on a Public One Way Encryption Function,"* U.S. Patent Number 4,908,861, March 13, 1990.

15. L. Brown, J. Pieprzyk, and J. Seberry, "LOKI – a cryptographic primitive for authentication and secrecy applications," *Advances in Cryptology, Proc. Auscrypt'90, LNCS 453*, J. Seberry and J. Pieprzyk, Eds., Springer-Verlag, 1990, pp. 229–236.

16. P. Camion, "Can a fast signature scheme without secret be secure ?" *Proc. 2nd International Conference on Applied Algebra, Algebraic Algorithms, and Error-Correcting Codes, LNCS 228*, A. Poli, Ed., Springer-Verlag, 1986, pp. 215–241.

17. P. Camion and J. Patarin, "The knapsack hash function proposed at Crypto'89 can be broken," *Advances in Cryptology, Proc. Eurocrypt'91, LNCS 547*, D.W. Davies, Ed., Springer-Verlag, 1991, pp. 39–53.

18. J.L. Carter and M.N. Wegman, "Universal classes of hash functions," *Proc. 9th ACM Symposium on the Theory of Computing*, 1977, pp. 106–112.

19. J.L. Carter and M.N. Wegman, "Universal classes of hash functions," *Journal of Computer and System Sciences*, Vol. 18, 1979, pp. 143–154.

20. C.C.I.T.T. X.509, *"The Directory — Authentication Framework,"* Recommendation, 1988, (same as ISO/IEC 9594-8, 1989).

21. D. Chaum and S. Roijakkers, "Unconditionally-secure digital signatures," *Advances in Cryptology, Proc. Crypto'90, LNCS 537*, S. Vanstone, Ed., Springer-Verlag, 1991, pp. 206–214.

22. D. Chaum, M. van der Ham, and B. den Boer, "A provably secure and efficient message authentication scheme," preprint, 1992.

23. F. Cohen, "A cryptographic checksum for integrity protection," *Computers & Security*, Vol. 6, 1987, pp. 505–510.

24. F. Cohen, *"The ASP integrity toolkit. Version 3.5,"* ASP Press, Pittsburgh (PA), 1991.

25. D. Coppersmith, "Another birthday attack," *Advances in Cryptology, Proc. Crypto'85, LNCS 218*, H.C. Williams, Ed., Springer-Verlag, 1985, pp. 14–17.

26. D. Coppersmith, "Analysis of ISO/CCITT Document X.509 Annex D," *IBM T.J. Watson Center, Yorktown Heights, N.Y., 10598, Internal Memo*, June 11, 1989, (also ISO/IEC JTC1/SC20/WG2/N160).

27. D. Coppersmith, "Two broken hash functions," *IBM T.J. Watson Center, Yorktown Heights, N.Y., 10598, Research Report RC 18397*, October 6, 1992.

28. J. Daemen, R. Govaerts, and J. Vandewalle, "A framework for the design of one-way hash functions including cryptanalysis of Damgård's one-way function based on a cellular automaton," *Advances in Cryptology, Proc. Asiacrypt'91, LNCS*, Springer-Verlag, to appear.

29. J. Daemen, A. Bosselaers, R. Govaerts, and J. Vandewalle, "Collisions for Schnorr's FFT-hash," *Presented at the rump session of Asiacrypt'91*.

30. J. Daemen, R. Govaerts, and J. Vandewalle, "A hardware design model for cryptographic algorithms," *Computer Security – ESORICS 92, Proc. Second European Symposium on Research in Computer Security, LNCS 648*, Y. Deswarte, G. Eizenberg, and J.-J. Quisquater, Eds., Springer-Verlag, 1992, pp. 419–434.

31. I.B. Damgård, "Collision free hash functions and public key signature schemes," *Advances in Cryptology, Proc. Eurocrypt'87, LNCS 304*, D. Chaum and W.L. Price, Eds., Springer-Verlag, 1988, pp. 203–216.

32. I.B. Damgård, "The application of claw free functions in cryptography," *PhD Thesis*, Aarhus University, Mathematical Institute, 1988.
33. I.B. Damgård, "A design principle for hash functions," *Advances in Cryptology, Proc. Crypto'89, LNCS 435*, G. Brassard, Ed., Springer-Verlag, 1990, pp. 416–427.
34. I.B. Damgård and L.R. Knudsen, "Some attacks on the ARL hash function," *Presented at the rump session of Auscrypt'92*.
35. D. Davies and W. L. Price, "The application of digital signatures based on public key cryptosystems," *NPL Report* DNACS 39/80, December 1980.
36. D. Davies, "A message authenticator algorithm suitable for a mainframe computer," *Advances in Cryptology, Proc. Crypto'84, LNCS 196*, G.R. Blakley and D. Chaum, Eds., Springer-Verlag, 1985, pp. 393–400.
37. D. Davies and W. L. Price, "Digital signatures, an update," *Proc. 5th International Conference on Computer Communication*, October 1984, pp. 845–849.
38. D. Davies and W.L. Price, *"Security for Computer Networks: an Introduction to Data Security in Teleprocessing and Electronic Funds Transfer (2nd edition),"* Wiley & Sons, 1989.
39. B. den Boer and A. Bosselaers, "An attack on the last two rounds of MD4," *Advances in Cryptology, Proc. Crypto'91, LNCS 576*, J. Feigenbaum, Ed., Springer-Verlag, 1992, pp. 194–203.
40. B. den Boer, personal communication.
41. B. den Boer and A. Bosselaers, "Collisions for the compression function of MD5," preprint, April 1992.
42. B. den Boer, "A simple and key-economical authentication scheme," preprint, 1992.
43. Y. Desmedt, "Unconditionally secure authentication schemes and practical and theoretical consequences," *Advances in Cryptology, Proc. Crypto'85, LNCS 218*, H.C. Williams, Ed., Springer-Verlag, 1985, pp. 42–55.
44. Y. Desmedt, "What happened with knapsack cryptographic schemes," in *"Performance Limits in Communication, Theory and Practice,"* J.K. Skwirzynski, Ed., Kluwer, 1988, pp. 113–134.
45. M. Desoete, K. Vedder, and M. Walker, "Cartesian authentication schemes," *Advances in Cryptology, Proc. Eurocrypt'89, LNCS 434*, J.-J. Quisquater and J. Vandewalle, Eds., Springer-Verlag, 1990, pp. 476–490.
46. W. Diffie and M.E. Hellman, "New directions in cryptography," *IEEE Trans. on Information Theory*, Vol. IT–22, No. 6, 1976, pp. 644–654.
47. T. ElGamal, "A public key cryptosystem and a signature scheme based on discrete logarithms," *IEEE Trans. on Information Theory*, Vol. IT–31, No. 4, 1985, pp. 469–472.
48. J.H. Evertse and E. Van Heyst, "Which new RSA-signatures can be computed from certain given RSA-signatures?" *Journal of Cryptology*, Vol. 5, No. 1, 1992, pp. 41–52.
49. V. Fåk, "Repeated uses of codes which detect deception," *IEEE Trans. on Information Theory*, Vol. IT–25, No. 2, 1979, pp. 233–234.
50. U. Feige, A. Fiat, and A. Shamir, "Zero knowledge proofs of identity," *Journal of Cryptology*, Vol. 1, No. 2, 1988, pp. 77–94.
51. FIPS 46, *"Data Encryption Standard,"* Federal Information Processing Standard, National Bureau of Standards, U.S. Department of Commerce, Washington D.C., January 1977.

52. FIPS 81, *"DES Modes of Operation,"* Federal Information Processing Standard, National Bureau of Standards, US Department of Commerce, Washington D.C., December 1980.

53. FIPS 113, *"Computer Data Authentication,"* Federal Information Processing Standard, National Bureau of Standards, US Department of Commerce, Washington D.C., May 1985.

54. FIPS xxx, *"Digital Signature Standard,"* Federal Information Processing Standard, Draft, National Institute of Standards and Technology, US Department of Commerce, Washington D.C., August 30, 1991.

55. FIPS yyy, *"Secure Hash Standard,"* Federal Information Processing Standard, Draft, National Institute of Standards and Technology, US Department of Commerce, Washington D.C., January 31, 1992.

56. A. Fujioka, T. Okamoto, and S. Miyaguchi, "ESIGN: an efficient digital signature implementation for smart cards," *Advances in Cryptology, Proc. Eurocrypt'91, LNCS 547*, D.W. Davies, Ed., Springer-Verlag, 1991, pp. 446–457.

57. E. Gilbert, F. MacWilliams, and N. Sloane, "Codes which detect deception," *Bell System Technical Journal*, Vol. 53, No. 3, 1974, pp. 405–424.

58. J.K. Gibson, "Discrete logarithm hash function that is collision free and one way," *IEE Proceedings-E*, Vol. 138, No. 6, November 1991, pp. 407–410.

59. M. Girault, "Hash-functions using modulo-n operations," *Advances in Cryptology, Proc. Eurocrypt'87, LNCS 304*, D. Chaum and W.L. Price, Eds., Springer-Verlag, 1988, pp. 217–226.

60. M. Girault, R. Cohen, and M. Campana, "A generalized birthday attack," *Advances in Cryptology, Proc. Eurocrypt'88, LNCS 330*, C.G. Günther, Ed., Springer-Verlag, 1988, pp. 129–156.

61. Ph. Godlewski and P. Camion, "Manipulations and errors, detection and localization," *Advances in Cryptology, Proc. Eurocrypt'88, LNCS 330*, C.G. Günther, Ed., Springer-Verlag, 1988, pp. 97–106.

62. S. Goldwasser, S. Micali, and R.L. Rivest, "A digital signature scheme secure against adaptive chosen-message attacks," *SIAM Journal on Computing*, Vol. 17, No. 2, 1988, pp. 281-308.

63. J.A. Gordon, "How to forge RSA certificates," *Electronic Letters*, Vol. 21, No. 9, 1985, pp. 377–379.

64. L.C. Guillou, M. Davio, and J.-J. Quisquater, "Public-key techniques: randomness and redundancy," *Cryptologia*, Vol. 13, April 1989, pp. 167–189.

65. L.C. Guillou, J.-J. Quisquater, M. Walker, P. Landrock, and C. Shaer, "Precautions taken against various potential attacks in ISO/IEC DIS 9796," *Advances in Cryptology, Proc. Eurocrypt'90, LNCS 473*, I.B. Damgård, Ed., Springer-Verlag, 1991, pp. 465–473.

66. G. Harper, A. Menezes, and S. Vanstone, "Public-key cryptosystems with very small key lengths," *Advances in Cryptology, Proc. Eurocrypt'92, LNCS 658*, R.A. Rueppel, Ed., Springer-Verlag, 1993, pp. 163–173.

67. F. Heider, D. Kraus, and M. Welschenbach, "Some preliminary remarks on the Decimal Shift and Add algorithm (DSA)," *Abstracts Eurocrypt'86, May 20-22, 1986, Linköping, Sweden*, p. 1.2. (Full paper available from the authors.)

68. M. Hellman, R. Merkle, R. Schroeppel, L. Washington, W. Diffie, S. Pohlig, and P. Schweitzer, "Results of an initial attempt to cryptanalyze the NBS Data Encryption Standard," *Information Systems Lab., Dept. of Electrical Eng., Stanford Univ.*, 1976.

69. Y.J. Huang and F. Cohen, "Some weak points of one fast cryptographic checksum algorithm and its improvement," *Computers & Security*, Vol. 7, 1988, pp. 503-505.

70. R. Impagliazzo and M. Naor, "Efficient cryptographic schemes provably as secure as subset sum," *Proc. 30th IEEE Symposium on Foundations of Computer Science*, 1989, pp. 236–241.

71. ISO 7498-2, *"Information processing – Open systems interconnection – Basic reference model – Part 2: Security architecture,"* ISO/IEC, 1987.

72. ISO 8730, *"Banking - Requirements for message authentication (wholesale),"* ISO, 1990.

73. ISO 8731, *"Banking - approved algorithms for message authentication - Part 1: DEA,"* ISO, 1987. *"Part 2, Message Authentication Algorithm (MAA),"* ISO, 1987.

74. ISO/IEC 9796, *"Information technology - Security techniques - Digital signature scheme giving message recovery,"* ISO/IEC, 1991.

75. ISO/IEC 9797, *"Information technology - Data cryptographic techniques - Data integrity mechanisms using a cryptographic check function employing a block cipher algorithm,"* ISO/IEC, 1989.

76. ISO/IEC 10116, *"Information technology - Security techniques - Modes of operation of an n-bit block cipher algorithm,"* ISO/IEC, 1991.

77. ISO/IEC DIS 10118, *"Information technology - Security techniques - Hash-functions - Part 1: General and Part 2: Hash-functions using an n-bit block cipher algorithm,"* ISO/IEC, 1992.

78. *"Hash functions using a pseudo random algorithm,"* ISO-IEC/JTC1/SC27/WG2 N98, Japanese contribution, 1991.

79. *"AR fingerprint function,"* ISO-IEC/JTC1/SC27/WG2 N179, working document, 1992.

80. R.R. Jueneman, S.M. Matyas, and C.H. Meyer, "Message authentication with Manipulation Detection Codes," *Proc. 1983 IEEE Symposium on Security and Privacy*, 1984, pp. 33-54.

81. R.R. Jueneman, S.M. Matyas, and C.H. Meyer, "Message authentication," *IEEE Communications Mag.*, Vol. 23, No. 9, 1985, pp. 29-40.

82. R.R. Jueneman, "A high speed Manipulation Detection Code," *Advances in Cryptology, Proc. Crypto'86, LNCS 263*, A.M. Odlyzko, Ed., Springer-Verlag, 1987, pp. 327–347.

83. R.R. Jueneman, "Electronic document authentication," *IEEE Network Mag.*, Vol. 1, No. 2, 1987, pp. 17-23.

84. A. Jung, "Implementing the RSA cryptosystem," *Computers & Security*, Vol. 6, 1987, pp. 342–350.

85. A. Jung, "The strength of the ISO/CCITT hash function," preprint, October 1990.

86. B.S. Kaliski, "The MD2 Message-Digest algorithm," *Request for Comments (RFC) 1319*, Internet Activities Board, Internet Privacy Task Force, April 1992.

87. X. Lai and J.L. Massey, "Hash functions based on block ciphers," *Advances in Cryptology, Proc. Eurocrypt'92, LNCS 658*, R.A. Rueppel, Ed., Springer-Verlag, 1993, pp. 55–70.

88. X. Lai, *"On the Design and Security of Block Ciphers,"* ETH Series in Information Processing, Vol. 1, J. Massey, Ed., Hartung-Gorre Verlag, Konstanz, 1992.

89. X. Lai, R.A. Rueppel, and J. Woollven, "A fast cryptographic checksum algorithm based on stream ciphers," *Advances in Cryptology, Proc. Auscrypt'92, LNCS*, Springer-Verlag, to appear.

90. C. Linden and H. Block, "Sealing electronic money in Sweden," *Computers & Security*, Vol. 1, No. 3, 1982, p. 226.

91. J.L. Massey, "Cryptography — A selective survey," *Digital Communications (Proc. 1985 International Tirrenia Workshop)*, E. Biglieri and G. Prati, Eds., Elsevier Science Publ., 1986, pp. 3-25.

92. J.L. Massey, "An introduction to contemporary cryptology," in *"Contemporary Cryptology: The Science of Information Integrity,"* G.J. Simmons, Ed., IEEE Press, 1991, pp. 3-39.

93. S.M. Matyas, C.H. Meyer, and J. Oseas, "Generating strong one-way functions with cryptographic algorithm," *IBM Techn. Disclosure Bull.*, Vol. 27, No. 10A, 1985, pp. 5658-5659.

94. R. Merkle and M. Hellman, "Hiding information and signatures in trapdoor knapsacks," *IEEE Trans. on Information Theory*, Vol. IT-24, No. 5, 1978, pp. 525-530.

95. R. Merkle, *"Secrecy, Authentication, and Public Key Systems,"* UMI Research Press, 1979.

96. R. Merkle, "A certified digital signature," *Advances in Cryptology, Proc. Crypto'89, LNCS 435*, G. Brassard, Ed., Springer-Verlag, 1990, pp. 218-238.

97. R. Merkle, "One way hash functions and DES," *Advances in Cryptology, Proc. Crypto'89, LNCS 435*, G. Brassard, Ed., Springer-Verlag, 1990, pp. 428-446.

98. R. Merkle, "A fast software one-way hash function," *Journal of Cryptology*, Vol. 3, No. 1, 1990, pp. 43-58.

99. C.H. Meyer and S.M. Matyas, *"Cryptography: a New Dimension in Data Security,"* Wiley & Sons, 1982.

100. C.H. Meyer and M. Schilling, "Secure program load with Manipulation Detection Code," *Proc. Securicom 1988*, pp. 111-130.

101. C. Mitchell, "Multi-destination secure electronic mail," *The Computer Journal*, Vol. 32, No. 1, 1989, pp. 13-15.

102. C. Mitchell, F. Piper, and P. Wild, "Digital signatures," in *"Contemporary Cryptology: The Science of Information Integrity,"* G.J. Simmons, Ed., IEEE Press, 1991, pp. 325-378.

103. S. Miyaguchi, M. Iwata, and K. Ohta, "New 128-bit hash function," *Proc. 4th International Joint Workshop on Computer Communications*, Tokyo, Japan, July 13-15, 1989, pp. 279-288.

104. S. Miyaguchi, "The FEAL cipher family," *Advances in Cryptology, Proc. Crypto'90, LNCS 537*, S. Vanstone, Ed., Springer-Verlag, 1991, pp. 627-638.

105. S. Miyaguchi, K. Ohta, and M. Iwata, "128-bit hash function (N-hash)," *Proc. Securicom 1990*, pp. 127-137.

106. J.H. Moore and G.J. Simmons, "Cycle structure of the DES for keys having palindromic (or antipalindromic) sequences of round keys," *IEEE Trans. on Software Engineering*, Vol. 13, 1987, pp. 262-273.

107. J.H. Moore, "Protocol failures in cryptosystems," in *"Contemporary Cryptology: The Science of Information Integrity,"* G.J. Simmons, Ed., IEEE Press, 1991, pp. 543-558.

108. M. Naor and M. Yung, "Universal one-way hash functions and their cryptographic applications," *Proc. 21st ACM Symposium on the Theory of Computing*, 1990, pp. 387-394.

109. T. Okamoto and K. Ohta, "A modification of the Fiat-Shamir scheme," *Advances in Cryptology, Proc. Crypto'88, LNCS 403*, S. Goldwasser, Ed., Springer-Verlag, 1990, pp. 232-243.

110. T. Okamoto, "Provably secure and practical identification schemes and corresponding signature schemes," *Advances in Cryptology, Proc. Crypto'92, LNCS,* E.F. Brickell, Ed., Springer-Verlag, to appear.

111. T. Okamoto and K. Ohta, "Survey of digital signature schemes," *Proc. of the 3rd symposium on State and Progress of Research in Cryptography,* W. Wolfowicz, Ed., Fondazione Ugo Bordoni, 1993, pp. 17–29.

112. J.C. Pailles and M. Girault, "The security processor CRIPT," *4th IFIP SEC,* Monte-Carlo, December 1986, pp. 127–139.

113. B. Preneel, A. Bosselaers, R. Govaerts, and J. Vandewalle, " A chosen text attack on the modified cryptographic checksum algorithm of Cohen and Huang," *Advances in Cryptology, Proc. Crypto'89, LNCS 435,* G. Brassard, Ed., Springer-Verlag, 1990, pp. 154–163.

114. B. Preneel, A. Bosselaers, R. Govaerts, and J. Vandewalle, "Collision free hash functions based on blockcipher algorithms," *Proc. 1989 International Carnahan Conference on Security Technology,* pp. 203-210.

115. B. Preneel, R. Govaerts, and J. Vandewalle, "Cryptographically secure hash functions: an overview," *ESAT Internal Report, K.U. Leuven,* 1989.

116. B. Preneel, A. Bosselaers, R. Govaerts, and J. Vandewalle, "Cryptanalysis of a fast cryptographic checksum algorithm," *Computers & Security,* Vol. 9, 1990, pp. 257–262.

117. B. Preneel, R. Govaerts, and J. Vandewalle, "On the power of memory in the design of collision resistant hash functions," *Advances in Cryptology, Proc. Auscrypt'92, LNCS,* Springer-Verlag, to appear.

118. B. Preneel, R. Govaerts, and J. Vandewalle, "An attack on two hash functions by Zheng, Matsumoto, and Imai," *Presented at the rump session of Auscrypt'92.*

119. B. Preneel, "Analysis and design of cryptographic hash functions," *Doctoral Dissertation,* Katholieke Universiteit Leuven, 1993.

120. J.-J. Quisquater and L. Guillou, "A "paradoxical" identity-based signature scheme resulting from zero-knowledge," *Advances in Cryptology, Proc. Crypto'88, LNCS 403,* S. Goldwasser, Ed., Springer-Verlag, 1990, pp. 216–231.

121. J.-J. Quisquater and J.-P. Delescaille, "How easy is collision search ? Application to DES," *Advances in Cryptology, Proc. Eurocrypt'89, LNCS 434,* J.-J. Quisquater and J. Vandewalle, Eds., Springer-Verlag, 1990, pp. 429–434.

122. J.-J. Quisquater and M. Girault, "2n-bit hash-functions using n-bit symmetric block cipher algorithms," *Abstracts Eurocrypt'89, April 10-13, 1989, Houthalen, Belgium.*

123. J.-J. Quisquater and M. Girault, "2n-bit hash-functions using n-bit symmetric block cipher algorithms," *Advances in Cryptology, Proc. Eurocrypt'89, LNCS 434,* J.-J. Quisquater and J. Vandewalle, Eds., Springer-Verlag, 1990, pp. 102–109.

124. J.-J. Quisquater and J.-P. Delescaille, "How easy is collision search. New results and applications to DES," *Advances in Cryptology, Proc. Crypto'89, LNCS 435,* G. Brassard, Ed., Springer-Verlag, 1990, pp. 408–413.

125. M.O. Rabin, "Digitalized signatures," in *"Foundations of Secure Computation,"* R. Lipton and R. DeMillo, Eds., Academic Press, New York, 1978, pp. 155-166.

126. M.O. Rabin, "Digitalized signatures and public-key functions as intractable as factorization," *Technical Report MIT/LCS/TR-212,* Massachusetts Institute of Technology, Laboratory for Computer Science, Cambridge, MA, January 1979.

127. "Race Integrity Primitives Evaluation (RIPE): final report," *CWI Report CS-R9324,* RACE 1040, 1993.

128. R.L. Rivest, A. Shamir, and L. Adleman, "A method for obtaining digital signatures and public-key cryptosystems," *Communications ACM*, Vol. 21, February 1978, pp. 120–126.

129. R.L. Rivest, "The MD4 message digest algorithm," *Advances in Cryptology, Proc. Crypto'90, LNCS 537*, S. Vanstone, Ed., Springer-Verlag, 1991, pp. 303–311.

130. R.L. Rivest, "The MD4 message-digest algorithm," *Request for Comments (RFC) 1320*, Internet Activities Board, Internet Privacy Task Force, April 1992.

131. R.L. Rivest, "The MD5 message-digest algorithm," *Request for Comments (RFC) 1321*, Internet Activities Board, Internet Privacy Task Force, April 1992.

132. J. Rompel, "One-way functions are necessary and sufficient for secure signatures," *Proc. 22nd ACM Symposium on the Theory of Computing*, 1990, pp. 387–394.

133. R.A. Rueppel, "Stream ciphers," in *"Contemporary Cryptology: The Science of Information Integrity,"* G.J. Simmons, Ed., IEEE Press, 1991, pp. 65–134.

134. C.P. Schnorr, "Efficient identification and signatures for smart cards," *Advances in Cryptology, Proc. Crypto'89, LNCS 435*, G. Brassard, Ed., Springer-Verlag, 1990, pp. 239–252.

135. C.P. Schnorr, "An efficient cryptographic hash function," *Presented at the rump session of Crypto'91*.

136. C.P. Schnorr, "FFT-Hash II, efficient cryptographic hashing," *Advances in Cryptology, Proc. Eurocrypt'92, LNCS 658*, R.A. Rueppel, Ed., Springer-Verlag, 1993, pp. 45–54.

137. C.E. Shannon, "Communication theory of secrecy systems," *Bell System Technical Journal*, Vol. 28, 1949, pp. 656-715.

138. G.J. Simmons, "A natural taxonomy for digital information authentication schemes," *Advances in Cryptology, Proc. Crypto'87, LNCS 293*, C. Pomerance, Ed., Springer-Verlag, 1988, pp. 269–288.

139. G.J. Simmons, "A survey of information authentication," in *"Contemporary Cryptology: The Science of Information Integrity,"* G.J. Simmons, Ed., IEEE Press, 1991, pp. 381–419.

140. D.R. Stinson, "Combinatorial characterizations of authentication codes," *Advances in Cryptology, Proc. Crypto'91, LNCS 576*, J. Feigenbaum, Ed., Springer-Verlag, 1992, pp. 62–73.

141. D.R. Stinson, "Universal hashing and authentication codes," *Advances in Cryptology, Proc. Crypto'91, LNCS 576*, J. Feigenbaum, Ed., Springer-Verlag, 1992, pp. 74–85.

142. S. Vaudenay, "FFT-hash-II is not yet collision-free," *Advances in Cryptology, Proc. Crypto'92, LNCS*, E.F. Brickell, Ed., Springer-Verlag, to appear.

143. G.S. Vernam, "Cipher printing telegraph system for secret wire and radio telegraph communications," *Journal American Institute of Electrical Engineers*, Vol. XLV, 1926, pp. 109-115.

144. R.S. Winternitz, "Producing a one-way hash function from DES," *Advances in Cryptology, Proc. Crypto'83*, D. Chaum, Ed., Plenum Press, New York, 1984, pp. 203–207.

145. A.C. Yao, "Theory and applications of trapdoor functions," *Proc. 23rd IEEE Symposium on Foundations of Computer Science*, 1982, pp. 80–91.

146. G. Yuval, "How to swindle Rabin," *Cryptologia*, Vol. 3, 1979, pp. 187–189.

147. G. Zémor, "Hash functions and graphs with large girths," *Advances in Cryptology, Proc. Eurocrypt'91, LNCS 547*, D.W. Davies, Ed., Springer-Verlag, 1991, pp. 508–511.

148. Y. Zheng, T. Matsumoto, and H. Imai, "Connections between several versions of one-way hash functions," *Proc. SCIS90, The 1990 Symposium on Cryptography and Information Security*, Nihondaira, Japan, Jan. 31–Feb.2, 1990.

149. Y. Zheng, T. Matsumoto, and H. Imai, "Duality between two cryptographic primitives," *Proc. 8th International Conference on Applied Algebra, Algebraic Algorithms and Error-Correcting Codes, LNCS 508*, S. Sakata, Ed., Springer-Verlag, 1991, pp. 379–390.

150. Y. Zheng, J. Pieprzyk, and J. Seberry, "HAVAL — a one-way hashing algorithm with variable length output," *Advances in Cryptology, Proc. Auscrypt'92, LNCS*, Springer-Verlag, to appear.

Key Management

Walter Fumy

Siemens AG, Dept. AUT 961

P.O. Box 3220, D-8520 Erlangen, Germany

Abstract

The purpose of key management is to provide secure procedures for handling cryptographic keying material to be used in symmetric or asymmetric cryptographic mechanisms. This includes user registration, key generation, key distribution, key storage, and key deletion. Key management schemes depend on the type of keys to be distributed, on the given facilities and on the specific application. For almost all systems, it is necessary to distribute keys over the same communication channels by which actual data are transmitted. Secure key distribution over such a channel requires cryptographic protection and thus the availability of matching keys. This circularity has to be broken through prior distribution of keys by different means.

1 Introduction

Security services based on cryptographic mechanisms (e.g. encipherment or authentication) assume cryptographic keys to be distributed to the communicating parties prior to secure communications. The secure management of these keys is one of the most critical elements when integrating cryptographic functions into a system, since even the most elaborated security concept will be ineffective if the key management is weak.

The main purpose of key management is to provide secure procedures for handling cryptographic keying material to be used in symmetric or asymmetric cryptographic mechanisms. This includes user registration, key generation, key distribution, key replacement, key storage, and key deletion. By the formalization of those procedures provision can be made for audit trails to be established. Part 2 of the Open Systems Interconnection (OSI) reference model, the security architecture, defines key management as "the generation, storage, distribution, deletion, archiving and application of keys in accordance with a security policy" [ISO 7498-2].

Key management schemes generally depend on the type of keys to be distributed, on the given facilities (e.g. the properties of the specific environment) and on the specific application. The most important considerations are the threats to be protected against, and the physical and architectural structure of the system. One central problem of key management is key distribution, i.e. the problem of establishing keying material whose origin, integrity, and - in the case of secret keys - confidentiality can be guaranteed. Another central aim of a key management system is to allow authentication of entities by means of keys.

Common cryptographic techniques used for the protection of data can also be used for the protection of keying material. However, many of the important properties of key management protocols do not depend on the underlying cryptographic algorithms, but rather on the structure of the messages exchanged. Therefore, bugs in such protocols usually do not come from weak cryptographic algorithms, but from mistakes in higher levels of the design.

A large variety of mechanisms for key establishment can be found in the literature. [DiHe 76] and [Ruep 88] e.g. describe procedures which allow the establishment of a common secret key for two users and which only require the communication of public messages. [Okam 86] proposes similar schemes that utilize each user's identification information to authenticate the exchanged messages. [Günt 89] and [BaKn 89] use data exchanged during an authentication protocol to construct a session key. [KoOh 87] shows how to generate a common secret conference key for two or more users that are connected in a ring, a complete graph, or a star network.

Key management has also been addressed by different standardization bodies which led to several national and international standards, in particular in the area of banking [ANSI X9.17] (see also [Bale 85]), [ISO 8731-1], [CCIT X.509]. Standards for other application areas as well as base standards dealing with generic issues of key management can be expected to follow [ISO 11666], [ISO KM-2], [ISO KM-3].

2 Key Management Security Issues

A fundamental security requirement for every key management system is the control of keying material through the entire lifetime of the keys in order to prevent unauthorized disclosure, modification, substitution, and replay. Cryptographic keys must be randomly chosen or generated. In order to protect (secret) keying material from disclosure it must either be physically secured or enciphered. In order to protect keying material from modification it must either be physically secured or authenticated. Authentication may involve the use of parameters like counters or timestamps to also protect from replay of old keys, insertion of false keys, and substitution or deletion of keys.

Keying material may be distributed either manually or automatically. Manually distributed keys are exchanged between parties by use of a courier or some other physical means (e.g. sealed envelopes, tamper-proof devices) independent of the communication channel. Manual key distribution is time consuming and expensive. When using only symmetric cryptographic techniques at least the first key has to be manually exchanged between two parties in order to allow secure communications.

An automatic distribution of keys typically employs different types of messages. A transaction usually is initiated by requesting a key from some central facility (e.g. a Key Distribution Center), or from the entity a key is to be exchanged with. Cryptographic Service

Messages (CSMs) are exchanged between communicating parties for the transmission of keying material, or for authentication purposes. CSMs may contain keys, or other keying material, such as the distinguished names of entities, key-IDs, counters, or random values. CSMs have to be protected depending on their contents and on the security requirements. Generic requirements include:

- **Data Confidentiality:** Secret keys and possibly other data are to be kept confidential while being transmitted or stored.

- **Modification Detection** is to counter the active threat of unauthorized modification of data items. In most environments all cryptographic service messages have to be protected against modification.

- **Replay Detection / Timeliness:** Replay detection is to counter unauthorized duplication of data items. Timeliness requires that the response to a challenge message is prompt and does not allow for play-back of some authentic response message by an impersonator.

- **Entity Authentication** is to corroborate that an entity is the one claimed.

- **Data Origin Authentication** is to corroborate that the source of a message is the one claimed. As defined in [ISO 7498-2] it does not provide protection against duplication or modification of the message. In practice, however, data origin authentication often is a combination of sender authentication and modification detection.

- **Proof of Delivery** shows the sender of a message that the message has been received by its legitimate receiver correctly.

For almost all systems, it is necessary to devise means to distribute keys over the same communication channels by which actual data are transmitted. Unfortunately, secure key distribution over such a channel requires cryptographic protection and thus the availability of matching keys. This circularity has to be broken through prior distribution of a number of keys by different means. For this purpose, in general some **Trusted Party** is required. If the mechanisms employed are based on conventional or symmetric methods, this trusted party usually is called **Key Distribution Center.** If the mechanisms used are based on asymmetric methods the trusted party is called **Key Certification Center** or **Certification Authority.** The responsibilities of a trusted party may include

- to enforce the security policy of the system,
- to identify and register entities,
- to guarantee the integrity of the system (i.e. that the soft- and hardware modules do perform as claimed).

To summarize, a secure key management systems depends on

- the functionality of the security measures,
- the quality of the security measures (i.e. their cryptographic strength and the quality of their implementation),
- some amount of trust into security managers, builders of the system, etc.
- the physical security of some devices and/or communication channels.

3 Key Management Concepts

A **Key Management Facility** is a protected enclosure (e.g. a room or a cryptographic device) involved in key management procedures. Its contents has to be protected from unauthorized disclosure, modification, substitution, replay, insertion or deletion. Key management facilities include physically secure devices that contain cryptographic elements (cryptographic hardware, cryptographic software, keying material). Such a device, when operated in its intended manner and environment, cannot be successfully penetrated either to disclose all or part of the data resident within the device or to allow operational modifications to the device.

The integrity of a key management facility has to be ensured. Well known methods to gain confidence in the correct functioning of a facility include formal specification and verification methods, detection and logging of attempted attacks, and the construction of the system by trusted personnel in a secure environment. In addition, every key management system will ultimately rely on some form of physical security and on the trustworthiness of the personnel operating it.

The latter problem may be eased by the division of responsibilities so that no single person can gain complete information about important material (e.g. complete copies of a master key). The process of utilizing two or more separate entities (usually persons), operating in concert, to protect sensitive information by splitting the knowledge of that information between those entities is called **Dual Control**.

Manually distributed keys shall always be protected using physical security or dual control. Physical channels are typically used to achieve confidentiality and authentication jointly. Examples of physical channels for authentication only are newspapers or compact discs. One common method of dual control is to have at least two keys which are generated independently, stored and transported separately and XORed within a secure environment to produce the actual cryptographic key.

The danger of a key being compromised is considered to increase with the length of time it has been in operation and the amount of data it has been used to encipher. Therefore, keys have to be changed from time to time in order to maintain a high level of security. Unless efficient distribution procedures are applied, frequent key changes are likely to create an unacceptable overhead. The widely accepted solution to this problem is to use the communication channel itself for key distribution. Obviously, when being sent over an insecure channel the keys must themselves be encrypted. **Key-Enciphering Keys** (also called secondary keys) are used much less than the actual communications keys (also called primary keys) and for that reason need not be replaced as often as those.

The concept of using a key to protect other keys reduces the problem of providing protection for a large number of keys to that of protecting a small number of keys. Also, the introduction of a **Key Hierarchy** significantly reduces the number of keys that cannot be distributed automatically. Depending on the actual security policy, keys may be labelled according to their function so that their use can be reserved for that purpose. Keying material also should have a limited lifetime based on time, use, or other criteria.

Authentication of keying material is an essential security requirement. In the asymmetric case, **Certificates** are issued for authentication of public keys. A so-called credential contains the identifying data of an entity together with its public key and possibly other information (e.g. an expiration date). Credentials are rendered unforgeable by some certifying information (e.g. a digital signature provided by the key certification center), thus becoming certificates. The validity of a certificate can be verified using some public information (e.g. a public key of the key certification center) whose authenticity is provided by manual procedures. Since verifying the validity of a certificate can be carried out without the assistance of the certification authority, the trusted party is only needed for issueing the certificates.

In every key management system there is a kernel where **Physical Security** has to be assumed. This physically protected data usually includes so-called **Master Keys**, which are on the top level of a key hierarchy and therefore cannot be protected using other keys. Physical security measures will often be necessary to ensure complete protection of keying material. But since physical security is costly, attempts will be made to minimize the need for it by using cheaper techniques wherever possible.

The following design criteria for a key management system are considered to be important:

- Minimize the number and complexity of trusted mechanisms involved. Especially, minimize the involvement of central mechanisms.
- Minimize physical activity. E.g., the use of couriers should be kept at a minimum (i.e. non-existent if possible). This requirement also implies that for registration, entities should not have to travel far (for large systems this suggests a hierarchical approach).
- Minimize the need for physical security, e.g. the number and size of tamper resistant devices, or the number of secure channels required.
- Achieve maximum flexibility with regard to specific key distribution protocols.

4   User Registration

Any secure system ultimately requires a procedure by which an individual or a device is authenticated to the system. A key management system only makes sense, if it guarantees the link between an entity and its uniquely defined representing keys. This link may be achieved by cryptographic methods, so that registration of entities becomes an essential task for a key management system.

The registration of a subject (e.g. a person or a device) is to allow automatic identification of that subject in the sequel. There are several types of identification. An absolute identification is provided if a link between an ID (e.g. a distinguished name or a device-ID) and some physical representation of the identified subject (e.g. a person or a device) can be established. Some applications only require a relative identification, i.e. a procedure that re-identifies a subject known under some ID without linking it to another representation. An identification can be carried out manually or automatically. Absolute identification always requires at least one initial manual identification (e.g. by showing a passport, or a device-ID).

Mutual authentication usually is based on the exchange of certificates. An entity is represented by its credentials that have to be generated upon registration. Whenever an entity is registered, a certificate is issued as a proof of registration. This may involve various procedures, from an entry in a specific file to a signature by the certification authority on the credentials (e.g. as described in [CCIT X.509]). Further examples for registration procedures are the installation of device keys, and the personalization of smart cards.

5   Key Generation

Secure key generation implies the use of a random or pseudo-random process involving random seeds, which cannot be manipulated. Requirements are that certain elements of the key space are not more probable than others and that it is not possible for the unauthorized to gain any knowledge about the keys. In general, tamper resistancy or at least tamper detectability is necessary to protect the secret parameters from being tapped. Random encipherment keys and seeds are important to the security of a system. Simply selecting numbers for those parameters makes it easier for an intruder to compromise the system's security.

There are always some keys (e.g. master keys) that should be generated by using a (truely) random process like tossing a coin or throwing a dice. Regardless of the specific method used for key generation, any predictability must be avoided. Before using a coin or a dice for the purpose of key generation it should be verified that it is not severely biased. Common software for generating pseudo-random numbers available on many computer systems usually is too predictable to be used for this purpose. Also the use of random number tables cannot be recommended for generating keys.

6   Key Distribution - A Modular Approach

As a result of varied design decisions appropriate to different circumstances, a large variety of key distribution protocols exist (see e.g. [MNSS 87], [NeSc 78], [OtRe 87]). Therefore, there is a need to explicate key distribution protocols in a way that allows to understand which goals the different protocols achieve and on which assumptions they depend. Formal methods devoted to the analysis of (authentication) protocols have been developed by M.Burrows, M.Abadi and R.Needham [BAN 90]. Another methodology that has the potential to largely automate the design of security architectures by keeping track of security requirements and management complexity has been presented by R.Rueppel [Ruep 90].

In the following a modular system is described that can be used to transform cryptographic protocols into a generic form and that has proven to be useful in the analysis and the construction of such protocols [FuMu 90]. Each of the building blocks described below addresses one or more of the security requirements given in section 2.

(a) Encipherment: The confidentiality of a data item D can be ensured by enciphering D with an appropriate key K. Depending on whether a secret key algorithm or a public key algorithm is used for the enciphering process, D will be enciphered with a secret key K shared between the sender and the legitimate recipient of the message (building block a1), or with the legitimate recipient B's public key K_{Bp} (a2). Encipherment with the sender A's private key K_{As} may be used to authenticate the origin of data item D, or to identify A (a3). Encipherment with a secret key (a1, a3) provides modification detection if B has some means to check the validity of D (e.g. if B knows D beforehand, or if D contains suitable redundancy).

generic:			$eK(D)$
(a1)	A	→ B:	$eK_{AB}(D)$
(a2)	X	→ B:	$eK_{Bp}(D)$
(a3)	A	→ X:	$eK_{As}(D)$

(b) Modification Detection Codes: To detect a modification of a data item D one can add some redundancy that has to be calculated using a collision-free function, i.e. it must be computationally infeasible to find two different values of D that render the same result. Moreover, this process has to involve a secret parameter K in order to prevent forgery. Appropriate combination of K and D also allows for data origin authentication. Examples of suitable building blocks are message authentication codes as defined in [ISO 9797] (see b1), or hash-functions, often combined with encipherment (b2 through b5).

generic:			$D \parallel mdcK(D)$
(b1)	A	→ B:	$D \parallel macK_{AB}(D)$
(b2)	A	→ B:	$D \parallel eK_{AB}(h(D))$
(b3)	A	→ X:	$D \parallel eK_{As}(h(D))$
(b4)	A	→ B:	$D \parallel h(K_{AB} \parallel D)$
(b5)	A	→ B:	$eK_{AB}(D \parallel h(D))$

These building blocks enable the legitimate recipient to detect unauthorized modification of the transmitted data immediately after receipt. The correctness of distributed keying material can also be checked if the sender confirms his knowledge of the key in a second step (see section (d) below).

(c) Replay Detection Codes: To detect the replay of a message and to check its timeliness, some explicit or implicit challenge and response mechanism has to be used, since the recipient has to be able to decide on the acceptance. This paragraph only deals with implicit mechanisms; explicit challenge and response mechanisms are being dealt with in section (d) (see below). In most applications the inclusion of a replay detection code (e.g. a timestamp TD, a counter CT, or a random number R) will only make sense if it is protected by modification detection. If modification detection of the data item D is required, the concatenation of D and the rdc also has to be protected against separation.

(c1)	$D := D \parallel rdc$	$rdc \in \{CT, TD\}$

With symmetric cryptographic mechanisms key modification can be used to detect the replay of a message. Building block c2 combines (e.g. XORs) the secret key with an rdc (e.g. a counter CT, or a random number R). Key offsetting used to protect data enciphered for distribution is a special case of building block c2. In this process the key used for encipherment is XORed with a count value [ANSI X9.17].

(c2) $K_{AB} := f(K_{AB}, rdc)$ $rdc \in \{R, CT\}$

(d) Proof of Knowledge of a Key: Authentication can be implemented by showing knowledge of a secret (e.g. a secret key). Nevertheless, a building block that proves the knowledge of a key K can also be useful, when K is public. There are several ways for A to prove to B the knowledge of a key that are all based on the principle of challenge and response in order to prevent a replay attack. Depending on the challenge which may be a data item in cleartext or in ciphertext, A has to process the key K and the rdc in an appropriate way (e.g. by enciperment (see d1), or by calculating a message authentication code (see d2)), or A has to perform a deciphering operation (d4).

The challenge may explicitly be provided by B (e.g. a random number R) or implicitly be given by a synchronized parameter (e.g. a timestamp TD, or a counter CT). For some building blocks the latter case requires only one pass to proof knowledge of K; its tradeoff is the necessary synchronization. If B provides a challenge enciphered with a key K*, A has to apply the corresponding deciphering key K. In these cases the enciphered data item has to be unpredictable (e.g. a random number R, or a key K**).

			generic:	authK(A to B)	
(d1)	A	←	B:	rdc	obsolete if $rdc \in \{CT, TD\}$
	A	→	B:	eK(rdc)	
(d2)	A	←	B:	rdc	obsolete if $rdc \in \{CT, TD\}$
	A	→	B:	rdc ‖ macK(rdc)	
(d3)	A	←	B:	rdc	obsolete if $rdc \in \{CT, TD\}$
	A	→	B:	rdc ‖ h(K ‖ rdc)	
(d4)	A	←	B:	eK*(rdc)	rdc random value
	A	→	B:	rdc	
(d5)	A	←	B:	eK*(K**) ‖ rdc	rdc arbitrary
	A	→	B:	eK**(rdc)	
(d6)	A	←	B:	eK*(K**) ‖ rdc	rdc arbitrary
	A	→	B:	macK**(rdc)	
(d7)	A	←	B:	eK*(K**) ‖ rdc	rdc arbitrary
	A	→	B:	h(K** ‖ rdc)	

Building blocks d5 through d7 of course also confirm that A knows K** in addition to the deciphering key K.

(e) Composition Rules: Some composition rules extracted from the description of the above building blocks and their effects on different security requirements are summarized in the table below.

Building Block	Requirement			
	Confidentiality	Modification Detection	Replay Detection	Data Origin Authentication
A → B: $eK_{AB}(D)$	yes	only if recipient can check validity of D	$D := D \| rdc$	only if recipient can check validity of D
X → B: $eK_{Bp}(D)$	yes	no	$D := D \| rdc$	no
A → X: $eK_{As}(D)$	no	only if recipient can check validity of D	$D := D \| rdc$	only if recipient can check validity of D
A → B: $D \| macK_{AB}(D)$	no	yes	$D := D \| rdc$ or $K_{AB} := f(K_{AB}, rdc)$	yes
A → B: $D \| eK_{AB}(h(D))$	no	yes	$D := D \| rdc$ or $K_{AB} := f(K_{AB}, rdc)$	yes
A → X: $D \| eK_{As}(h(D))$	no	yes	$D := D \| rdc$	yes
A → B: $D \| h(K_{AB} \| D)$	no	yes	$D := D \| rdc$ or $K_{AB} := f(K_{AB}, rdc)$	yes
A → B: $eK_{AB}(D \| h(D))$	yes	yes	$D := D \| rdc$ or $K_{AB} := f(K_{AB}, rdc)$	yes

The building blocks (cx) have to be applied first, so that the combination of (c1) with (a1) e.g. results in

$$A \rightarrow B: \qquad eK_{AB}(D \| rdc)$$

Besides the above composition rules one also can give several general rules:

(e1) A secret key shall not be used for both encipherment and authentication of the same data item.

(e2) Two consecutive transmissions from A to B may be replaced by one transmission of concatenated messages.

(e3) $D1 \| D2$ may be replaced by $D2 \| D1$.

(e4) $eK(D1) \| eK(D2)$ may be replaced by $eK(D1 \| D2)$.

7 Point-to-Point Key Distribution

The basic mechanism of every key distribution scheme is point-to-point key distribution. If based on symmetric cryptographic techniques point-to-point key distribution requires that the two parties involved already share a key that can be used to protect the keying material to be distributed. If based on asymmetric techniques point-to-point key distribution requires that each of the two parties has a public key with its associated secret key, and the certificate of the public key produced by a certification authority known to the other party.

In this section we discuss protocols for point-to-point distribution of a secret key that are derived from [ISO 9798-2], [ANSI X9.17], and [ISO 9798-3], respectively. The first example is to illustrate the construction of a key distribution protocol out of building blocks. Generic descriptions are used to exhibit similarities and differences between the discussed protocols.

General assumptions are:
- The initiator A is able to generate or otherwise acquire a secret key K^*.
- Security requirements are confidentiality of K^*, modification and replay detection, mutual authentication of A and B, and a proof of delivery for K^*.

For point-to-point key distribution protocols based on symmetric cryptographic techniques we additionally assume:
- A key K_{AB} is already shared by A and B.

The first two security requirements can be met by an appropriate combination of building blocks a1 through c2 (see table 1), whereas for the other two requirements one can choose from building blocks d1 through d7. As a first example we show a protocol built up from building blocks a1 and c1 (step 1: confidentiality of K^*, modification and replay detection), d1 and d5 (steps 2 to 4: mutual proof of knowledge of K_{AB} that includes B's proof of knowledge of K^*), and d4 (steps 5 and 6: A's proof of knowledge of K^*).

A	Protocol 1: Point-to-Point Key Distribution	B
$(1) \rightarrow$	$eK_{AB}(K^* \| rdc_1)$	
$(2) \rightarrow$	$eK_{AB}(rdc_2)$	
$(3) \rightarrow$	$eK_{AB}(K^*) \| rdc_3$	
	$eK^*(rdc_3)$	$\leftarrow (4)$
	$eK^*(rdc_4)$	$\leftarrow (5)$
$(6) \rightarrow$	rdc_4	

The above protocol can be greatly simplified by identifying the parameters rdc_2 and rdc_3 with rdc_1 ($= N$ which can be chosen to be a counter CT or a timestamp TD), and by applying composition rules e2 and e4 to steps 4 and 5. The resulting point-to-point key distribution protocol is one proposed in [ISO KM-2].

A	Protocol 1': Point-to-Point Key Distribution (ISO/IEC CD 9798-2)	B
(1) →	$eK_{AB}(K^* \| N)$	
	$eK^*(N \| R)$	← (2)
(3) →	R	

In generic form this protocol can be desribed as follows:

A	Protocol 1': Generic Form	B
(1) →	$eK(K^* \| rdc)$	
(2) →	$authK^*(A \text{ to } B)$	
	$authK^*(B \text{ to } A)$	← (3)

A point-to-point key distribution protocol proposed in [ANSI X9.17] takes a somewhat different approach. To achieve replay and modification detection protocol 2 (see below) makes use of building blocks c1 and b1 (see also table 1). A proves to B its knowledge of K* (and thus the knowledge of K_{AB}) using building block d2, whereas B proves its knowledge of K_{AB} with building block d6 which also confirms the correct receipt of K*.

A	Protocol 2: Point-to-Point Key Distribution (ANSI X9.17)	B
(1) →	$D \| eK_{AB}(K^*) \| N \| macK^*(...)$	
	$macK^*(D)$	← (2)

The generic form of protocol 2 exhibits the essential differences between the two protocols.

A	Protocol 2: Generic Form	B
(1) →	$eK(K^*) \| rdc \| mdcK^*(...)$	
	$authK^*(B \text{ to } A)$	← (2)

Finally we give an example for a point-to-point key distribution protocol based on public key techniques. We make the following supplementary assumptions:
- There is no shared key known to A and B before the key exchange process starts.
- There is a trusted third party C, where A can receive a certificate that contains the distinguished names of A and C, A's public key K_{Ap}, and the certificate's

expiration date TE. The integrity of the certificate is protected by C's signature.

As an example A's certificate is shown below:

$$IDc \parallel ID_A \parallel K_{Ap} \parallel TE \parallel eK_{Cs}(h(IDc \parallel ID_A \parallel K_{Ap} \parallel TE))$$

The exchange of certificates can be performed off-line and is not shown in the following protocol. In this protocol A sends a message (often refered to as *token*) to B that consists of a secret key K* enciphered with B's public key (building block a2) and an appended rdc. The integrity of the token is protected by A's signature (building block b3 combined with c1). This guarantees modification and replay detection, as well as data origin authentication. B responds with the enciphered rdc thereby acknowledging that it has received the key K* (building block d3).

A	Protocol 3: Point-to-Point Key Distribution (ISO/IEC CD 9798-3)	B
(1) →	eK$_{Bp}$(K*) || rdc || eK$_{As}$(h(eK$_{Bp}$(K*) || rdc))	
	eK*(rdc)	← (2)

The generic form of protocol 3 shows its similarity with protocol 2.

A	Protocol 3: Generic Form	B
(1) →	eK(K*) || rdc || mdcK'(...)	
	authK*(B to A)	← (2)

8 Key Distribution Centers

One of the simplest forms of key distribution employs a single **Key Distribution Center** (KDC) or **Certification Authority** for the entire system. Obvious disadvantages of this approach are its effect on network reliability and that centralized key distribution facilities can easily become a bottleneck. One possibility to overcome this problem is the introduction of redundant key distribution centers.

For large systems, key distribution usually is organized in a hierarchical way. There will be "local" KDCs, "regional" KDCs, and a "global" KDC. A local KDC is able to communicate securely with its "local" entities, i.e. with entities that are registered at the local KDC. Such a group of entities is called a **Domain**.

For communicating parties belonging to the same domain, key distribution procedures are the same as for the centralized approach. For communicating parties belonging to different local KDCs, those local KDCs have to establish a secure channel using a KDC at some higher level. Due to the hierarchical structure of the KDCs, any two entities registered under different local KDCs have a unique common KDC as ancestor. This design obviously can be generalized to as many levels as desired.

A main purpose of a KDC is to generate or acquire and distribute keys to parties that each already share a key with the KDC. Another important purpose of a KDC often is key translation. There is a great variety of protocols for both purposes. The following example is a key translation protocol proposed in [ISO KM-2] which is derived from a standard authentication mechanism [ISO 9798-2].

In this case the originator A sends a message to the KDC containing the distinguished name ID_B of the recipient B, the key K* and a number N, which may be a random or non-repeating number (e.g. sequence number, time). The whole message is enciphered under K_{AC}, namely the key enciphering key shared between A and the KDC. When a reply is received from the KDC the number N will be used to determine that it is not the result of a replay. The KDC deciphers the received message, re-enciphers it with a key K_{BC} shared with the second party and sends the re-enciphered message back to the originator who forwards it to B. For the protocol shown below we assume:

-   There is no shared key known to A and B before the key exchange process begins.

-   There is a trusted third party, the Key Distribution Center C, with which both A and
    B can communicate securely using the key enciphering keys K_{AC} and K_{BC}
    respectively.

-   The initiator A shall be able to generate or otherwise acquire a key K*.

-   Security requirements are confidentiality of K*, modification and replay detection,
    mutual authentication of the communicating parties, and a proof of delivery for
    K*.

A		C		B
(1) →	$eK_{AC}(ID_B \| K* \| N)$			
	$eK_{AC}(eK_{BC}(ID_A \| K*) \| N)$	← (2)		
(3) →	$eK_{BC}(ID_A \| K*)$			
(4) →	rdc			
	$eK*(rdc)$			← (5)

The above protocol of course can be simplified by combining messages (3) and (4). If rdc is chosen to be a counter CT, or a timestamp TD, message (4) is superfluous.

In a typical key distribution protocol one of the parties (the originator) requests a key from the KDC for later communication to another party. The KDC generates or acquires the key and sends a CSM to the originator protected by a key shared with the originator. This message contains a second CSM protected by a key shared between the KDC and the second party. This second CSM is then sent by the originator to the ultimate recipient. Following the distribution of a key by the KDC, the two parties may operate in a point-to-point mode.

9   Key Maintenance

Key maintenance includes procedures for key activation, key storage, key replacement, key translation, key recovery, black listing of compromised keys, key deactivation, and key deletion. The most important issues of key maintenance are addressed below.

Storage of Keying Material. A key storage facility provides secure storage of keys, e.g. confidentiality and integrity for secret keying material, or integrity for public keys. Secret keying material must be protected by physical security (e.g. by storing it within a cryptographic device) or enciphered by keys that have physical security. For all keying material unauthorized modification must be detectable by suitable authentication mechanisms.

Key Derivation is a technique by which a (potentially large) number of keys are generated from a single seed key (the "derivation key") and some variable data (e.g. the device-IDs). This technique allows to separate keys without the need to manage all the keys separately. The generation of "derived keys" utilises a non-reversible process such that the compromise of a derived key does not disclose the seed key or any other derived keys. Key derivation can also be used to update keys and thus obtain new keys without the need for a key distribution process.

Key Replacement. A key shall be replaced when its compromise is known or suspected. A key shall also be replaced within the time deemed feasible to determine it by an exhaustive attack. A replaced key shall not be re-used. The replacement key shall not be a variant or any non-secret transformation of the original key.

Key Recovery. Cryptographic keys may become lost due to human error, software bugs, or hardware malfunction. In communication security a simple handshake at session initiation can ensure that both entities are using the same key. Also message authentication techniques can be used for testing that plaintext has been recovered using the proper key. Key authentication techniques permit keys to be validated prior to their use. In the case where a key was lost, it still may be possible to revover that key by searching part of the key space. This approach may be successful, if the number of likely candidates is small enough.

Key Deletion. Destroying a key means eliminating all records of this key, such that no information remaining after the deletion provides any feasibly usable information about the destroyed key. A key may be destroyed by over-writing it with a new key or by zeroising it. Keying material stored on magnetic media should either be zeroised or the media itself should be destroyed.

References:

[ANSI X9.17]   ANSI X9.17-1985: *Financial Institution Key Management (Wholesale)*, 1985.

[Bale 85]   Balenson, D.M.: "Automated Distribution of Cryptographic Keys Using the Financial Institution Key Management Standard", IEEE Communications Magazine, July 1985, 41-46.

[BaKn 89]   Bauspieß, F.; Knobloch, H.-J.: "How to Keep Authenticity Alive in a Computer Network", Proceedings of Eurocrypt'89, Springer LNCS 434 (1990), 38-46.

[BAN 90]   Burrows, M.; Abadi, M.; Needham, R.: *A Logic of Authentication*, DEC System Research Center Report 39, 1990.

[CCIT X.509]   CCITT Recommendation X.509: *The Directory - Authentication Framework*, 1989.

[DiHe 76]   Diffie, W.; Hellman, M.E.: "New Directions in Cryptography", IEEE Transactions on Information Theory, 22 (1976), 644-654.

[FuLa 90]   Fumy, W.; Landrock, P.: "Principles of Key Management", Proceedings of CS'90: Symposium on Computer Security, Fondazione Ugo Bordoni (1991), 122-132.

[FuLe 90]   Fumy, W.; Leclerc, M.: "Integration of Key Management Protocols into the OSI Architecture", Proceedings of CS'90: Symposium on Computer Security, Fondazione Ugo Bordoni (1991), 151-159.

[FuMu 90]   Fumy, W.; Munzert, M.: "A Modular Approach to Key Distribution", Proceedings of Crypto'90, Springer LNCS (1991).

[Günt 89]   Günther, Ch.G.: "An Identity-Based Key-Exchange protocol", Proceedings of Eurocrypt'89, Springer LNCS 434 (1990), 29-37.

[ISO 7498-2]   ISO International Standard 7498-2: *"Open Systems Interconnection Reference Model - Part 2: Security Architecture"*, 1988.

[ISO 8731-1]   ISO International Standard 8731-1: *"Banking - Approved Algorithms for Message Authentication Part 1: DEA"*, 1987.

[ISO 9797]   ISO/IEC International Standard 9797: *Data Integrity Mechanism Using a Cryptographic Check Function Employing a Block Cipher Algorithm*, 1989.

[ISO 9798-2]   ISO/IEC Committee Draft 9798-2: *Entity Authentication Mechanisms - Part 2: Entity Authentication Using Symmetric Techniques, 1990*.

[ISO 9798-3]   ISO/IEC Committee Draft 9798-3: *Entity Authentication Mechanisms - Part 3: Entity Authentication Using a Public-Key Algorithm*, 1991.

[ISO 11666]   ISO/IEC Committee Draft 11666: *Banking - Key Management by Means of Asymmetric Algorithms*, 1991.

[ISO KM-2]   ISO/IEC/JTC1/SC27/WG2 Working Draft: *Key Management Part 2: Key Management Using Symmetric Cryptographic Techniques*, 1990.

[ISO KM-3]   ISO/IEC/JTC1/SC27/WG2 Working Draft: *Key Management Part 3: Key Management Using Public Key Techniques*, 1990.

[KoOh 87]   Koyama K.; Ohta, K.: "Identity-Based Conference Key Distribution Systems", Proceedings of Crypto'87, Springer LNCS 293 (1988), 175-184.

[MNSS 87]   Miller, S.P.; Neuman, C.; Schiller, J.I.; Saltzer, J.H.: *Kerberos Authentication and Authorization System*, Project Athena Technical Plan, MIT, 1987.

[NeSc 78]   Needham, R.M.; Schroeder, M.D.: "Using Encryption for Authentication in
             Large Networks of Computers", Communications of the ACM, **21**
             (1978), 993-999.

[Okam 86]   Okamoto, E.: "Proposal for Identity-Based Key Distribution Systems",
             Electronic Letters, **22** (1986), 1283-1284.

[OtRe 87]   Otway, D.; Rees, O.: "Efficient and Timely Mutual Authentication", Operating
             Systems Review, **21** (1987), 8-10.

[RSA 78]   Rivest, R.L.; Shamir, A.; Adleman, L.: "A Method for Obtaining Digital
             Signatures and Public-Key Cryptosystems", Comm. of the ACM **21**
             (1978), 120-126.

[Ruep 88]   Rueppel, R.A.: "Key Agreements Based on Function Composition",
             Proceedings of Eurocrypt'88, Springer LNCS **330** (1988), 3-10.

[Ruep 90]   Rueppel, R.A.: "Security Management", Proceedings of CS'90: Symposium on
             Computer Security, Fondazione Ugo Bordoni (1991), 43-50.

Section 3

Applications

EVALUATION CRITERIA FOR IT SECURITY

D. W. ROBERTS
Data Sciences (UK) Ltd.

ABSTRACT

After an Introduction which gives an outline of the paper's scope the first main section considers the History of the development of the various criteria. This leads to a discussion of the Aims and Purposes of the various authors and sponsors in the main thrust of each of these documents; it is interesting and relevant to know what the authors thought they were doing. Clearly some of the criteria were written with very limited scope in view while others had wider briefs. With an understanding of the aims and purposes of the criteria and their authors and sponsors it is possible to assess the level of Achievement in the various programs; we can ask to what extent they fulfilled the purposes. One of the most interesting aspects of systems security is the balance between the needs of the various sectors; the traditional dichotomy of Civil versus Military is very pronounced in this field. A final section looks at the Prospects for the Future, both in terms of the standards-making bodies and in terms of user needs

THE AUTHOR

David Roberts has been active in computing for thirty years and in the security aspects since the mid-seventies. In recent years he has been one of the two consultants to the UK Department of Trade and Industry Commercial Computer Security Centre which produced the "Green Books" and an expert consultant for the various programs of the Commission of the European Communities, including TEDIS, ESPRIT and RACE. Prior to that, he had extensive commercial experience as Chief Designer (Trusted Computing) with Plessey, as a consultant for Leasco and Dataskil, and as EDP Manager for Strand Hotels where he was responsible for the world's first successful hotel computer system.

He is a member of the Scientific Committee of the Centre d'Enseignement et de Recherche Appliqués au Management advising on the content of the Mastère Spécialisé en Sécurité des Systèmes d'Information course; a member of the British Standards Institute Technical Experts Panel on Security Evaluation Criteria (IST33/-/P3) which is the BSI part of the International Standards Organisation Technical Experts Panel on Security Evaluation Criteria (JTC1/SC27/WG3). He is the author of the book *Computer Security: Policy, Planning and Practice* and has published many technical papers in various fields since 1967.

Introduction

This paper gives an overview of the current state-of-the-art in the field of computer security evaluation criteria with particular emphasis on the commercial and industrial sectors. It reviews the history, the present status and the current prospects in the field.

Security evaluation criteria are dealt with in this paper in terms of a brief look at the historical perspective and the degree to which the historical aims and purposes were achieved; this is followed by a comparison of today's needs for military and civil purposes together with some thoughts as to the prospects in the near future.

There is a problem of the language used to consider the field; the anglophone has two separate words *security* and *safety* which are not differentiated in any other European Community languages. This can lead to confusion where there is an insistence (as, for instance, promulgated by the UK Ministry of Defence) that subsystems for protecting the confidentiality, integrity and/or availability of information held within computers processing information *per se* (security) are quite distinct and different from subsystems for protecting the confidentiality, integrity and/or availability of information held within computers whose malfunctioning may endanger human lives (safety). In most other languages there is no such distinction and the author cannot see that there is any significant difference.

History

The early work in this field was almost exclusively military and concerned with secrecy, although commercial computer security papers were being given at conferences as early as 1977. No consideration of the history of the development of the various criteria could possibly fail to acknowledge the debt that all the various criteria owe to the early work of David Bell and Leonard La Padula done for the MITRE Corporation [BandP]. This seminal work formed the basis of most, if not all, serious efforts to create and evaluate security up to the mid-1980's and has greatly influenced all the modern work also. In this paper they set out the basis of the mathematical treatment of security labels and clearances and defined the concept of dominance. These were oriented towards the defence industry hierarchies of sensitivity (SECRET, CONFIDENTIAL, UNCLASSIFIED, etc.) and to non-hierarchical categories of information (ATOMIC, CRYPTO, etc.); they did not cover caveats (nationality of readership, etc.).

The earliest official criteria in the field were those developed in the early 1980's by the US Department of Defense National Computer Security Centre; they were published in 1983 and are universally called the Orange Book [TCSEC]. The theoretical foundations of the Orange Book were strongly based on Bell and La Padula [BandP] but it was concerned only marginally with integrity and availability. The Orange Book was developed with a specific need in mind; it is not surprising, therefore, that it reads like a statement of user requirement, nor that the Security Policy is inextricably included in the list of needs. This is at its most obvious when the linkage between the level of assurance and the functionality is considered - the more secure the system is, the more functions it is required to perform.

In the later 1980's the problems of security which had long concerned managers in the commercial field began to be acknowledged by the relevant government bodies. The German government published a set of criteria which had much in common with the Orange Book but which was specified as covering commercial as well as military security [ZSIEC]. The French government also produced some criteria, but these were never made available to the commercial sector [FREC]. The UK Department of Trade and Industry produced a set of draft criteria for the evaluation of commercial computer security products; this was paralleled by a set of criteria produced, partly in secret, for government systems by the Government's Communications-Electronics Security Group.

In 1988/9 the four European sets of documents were merged into a common set of criteria which were published as a draft in May 1990 [ITSEC]. This document claimed to cover systems and products for commercial and military applications; moreover, it introduced a new concept to security, that of ease of use. It claims that a system's security is in doubt if the security system is not designed with ease of use as a primary criterion.

Aims and Purposes

USA Criteria

The declared aims and purposes of the Orange Book [TCSEC] are three-fold:

- To provide users with a metric with which to evaluate the degree of trust that can be placed in computer systems for the secure processing of classified and other sensitive information.

- To provide guidance to manufacturers as to what security features to build into their new and planned, commercial products in order to provide widely available systems that satisfy trust requirements for sensitive applications.

- To provide a basis for specifying security requirements in acquisition specifications.

These aims are to be understood within the historical perspective of the work being undertaken under the aegis of the DoD Computer Security Center which intended them to be used in the evaluation and selection of ADP being considered for the processing and/or storage and retrieval sensitive or classified information by the Department of Defense.

German Criteria

The German Criteria [ZSIEC] were developed:

to serve as a basis for the evaluation of IT systems as regards their protection mechanisms and the extent to which they can be trusted.

This aim is clearly not limited only to confidentiality, nor exclusively to governmental systems. The generalisation to commercial systems is intended to be achieved by the disentangling of security functionality from assurance levels and the provision of Functionality Classes to describe systems of various types. The first five Functionality Classes correspond to the five significant levels of the Orange Book ITCSEC~ and the remaining five are concerned with databases, high availability systems, information integrity, cryptographic products, and systems using public networks.

UK Criteria

The UK DTI Green Books [DTIEC] were intended to specify:

a) Technical Criteria for the security evaluation of IT products.

b) an evaluation and certification scheme for the practical application of the Technical Criteria.

c) associated user and vendor codes of practice.

These documents were only issued at Draft status for industry comment but it was abundantly clear that they were welcomed as a real conceptual step forward. This is because they were not concerned with complete systems but rather with products which could be incorporated in systems to improve (or even provide) the security of those systems. It was clear from the various public statements about the remit of the CCSC that they were never intended for use with non-commercial systems.

The UK CESG documents [CESG3] are a combination of publicly available and, one suspects, unpublished ideas. The published element of this set of Criteria:

promulgates standard descriptions for a set of levels of confidence that will be used by the CESG in the certification of computer systems and components that are to be used to process, store or forward classified information.

This is to be understood within the context of the responsibility of CESG which is to advise the UK Government on technical aspects of electronic information security to help ensure the protection of classified information against unauthorised disclosure. Thus CESG is primarily concerned with confidentiality.

Pan-European Criteria

The snappily titled Information Technology Security Evaluation Criteria (Harmonised Criteria of France - Germany - the Netherlands - the United Kingdom) [ITSEC] was prepared with its aim:

limited to harmonisation of existing criteria,

qualified by the additional caveat that:

it has sometimes been necessary to extend what already existed.

Any judgement of the extent of harmonisation and the items added in the "necessary extension" is complicated by the fact that the French contribution is not available outside the French members of the drafting group, even to workers in the field, although it is referenced in the document [FREC]. Further complications arise from the fact that the Netherlands does not even quote a title for their contribution.

Achievements

USA Criteria

The achievements of the authors of the Orange Book [TCSEC] cannot be doubted; it is a superb statement of what security features to build into ... products in order to provide ... systems that satisfy trust requirements peculiar to the US DoD and specific to mainframe computers. There are now "interpretations" for use with networks, products and databases. The mistake that has been made, especially in the USA but also in Europe, is to attempt to force all systems and products in the commercial field to conform to it. The reason that these efforts fail is that the underlying security policy of the government is different from that implemented by most companies (see the next section). It also linked functionality inextricably with assurance, assuming that the highest levels of assurance would always be associated with the widest ranges of functionality.

German Criteria

The German Criteria [ZSIEC] achieve, at least in part, the aim of being suitable for products as well as systems, and for commercial as well as military applications. The separation of assurance levels from functionality allows for the evaluation of products and for different priorities and security policy models to be accommodated. Unfortunately, however clearly it is stated that the Functionality Classes are only examples, there is always an assumption that the ones given in the document are the only acceptable ones. There is no attempt to give a description of a well-formed Functionality Class, nor instructions as to how new ones can become "official".

UK Criteria

The The DTI Green Books [DTIEC] seem to achieve their stated aims, though there are some doubts about some aspects of those objectives. They have a similar separation to that of the German Criteria whereby the definition of the functionality of the product is entirely independent of the assurance level sought. Furthermore, unlike the [ZSIEC] there is complete freedom of expression of the functionality by means of a Claims Language which allows any functionality to be described. They envisage that recognised "bundles" of claims will arise which would be analogous to the functionality classes.

The doubts about this approach arise from the possibility of a product being evaluated (possibly to a high level of assurance) for a single, trivial, claim and then advertised as an evaluated product in a misleading fashion. This fear is well-founded when one considers the plethora of products on the US market, described as designed to Orange Book B3 Criteria but which are, in fact, totally unevaluated.

The UK Government criteria openly published by CESG [CESG3] consist of assurance levels expressed in fairly vague terms. One can only assume that these documents are the tip of the iceberg and that the further documentation (referenced in [CESG3]) contain more explicit definitions and functionality descriptions which are considered not suitable for general publication.

Pan-European Criteria

In the harmonisation efforts it is obvious that there has, indeed, been considerable compromise. At the level of general principles, the German Criteria [ZSIEC] and the two UK sets of criteria [CESG3] and [DTIEC] have combined reasonably smoothly. It seems a little odd that the Functionality Groups and the Claims language were both needed but as the originator of the Claims Language it appears that I may be biased. The latest version of the ITSEC, the revision of January 1991 was issued only to a limited number of reviewers and had addressed some of the problems raised by those who commented on the 6000 copies of the original version. Unfortunately, the extension to include at least rudimentary treatment of hardware and of encryption algorithms (which were the widest demanded changes) have not been made. Hardware has been declared

> beyond the scope of the document

while the strength of encryption algorithms will be

> the subject of certification by the appropriate national body.

The wholly inexplicable addition to the previously published work is the new requirement that a secure system/product is of necessity one which is designed with ease-of-use as a major goal. This suggestion has two major drawbacks: it is not true that a very complex system can always be easy to use or even can be designed with that as a major goal, and secondly that ease-of-use is a wholly subjective measure. As an example of this, the use of graphical interfaces with icons is the current fashion in HCI but there are a significant number of users who are conceptual (rather than visual) thinkers and they have difficulty remembering what the little pictures mean.

CIVIL versus MILITARY

Military information secrecy and commercial information integrity have, at first glance, little in common; it would seem surprising, therefore, that there are so many techniques considered applicable to both. However, the techniques used in each case, and in the case of safety-critical systems, are all in one of two main classes: ensuring the reliability of the system or ensuring the reliability of the data.

The first of these two areas, the reliability of systems, includes the use of high-quality software (possibly including formal methods) and of reliable and dependable hardware (non-stop systems and/or formally proven hardware). The second area, reliability of data, includes such activities as encryption and signature of information to forestall (or at least detect) illegal access and amendment.

In the commercial area the integrity of information is often of paramount importance, whereas the military field requires confidentiality as its first priority. It is when one considers these aspects that the anglophone's distinction between safety and security (incidentally not supported in any other European language) is seen to be almost irrelevant to the discussion of techniques and standards.

PROSPECTS FOR THE FUTURE

Forecasting future developments, in any field, is always a risky venture; however, this paper will attempt to do so. The current innovative efforts in the field of evaluation criteria can be divided into three main parts, the USA, ISO and Europe, the latter of which is (or appears to be) amphisbaenic.

In the USA the efforts of NIST (née NBS) are devoted to ever more tortuous *interpretations* of the Orange Book [TCSEC] to cover Networks, Products (or sub-systems), Databases and so on. These are causing immense difficulties of interpretation for three main reasons:

a)   the basic problems with the Orange Book's lack of distinction between functionality and assurance,

b)   the US Government's apparent wish to apply this standard to an ever increasing range of situations where the US Navy's Security Policy is inappropriate, and

c)   US industry's protectionist attempts to lock all NATO Governments (at least) into using this standard under which only one US Government agency can issue certificates and then only to US companies.

In the ISO forum the immensely ponderous wheels which lead to an international standard have started to turn. A joint committee with the International Electrotechnical Commission [IEC] has been set up, known by the catchy title ISO/IEC JTC1 "Information Technology"; below this there are a large number of subcommittees of which ISO/IEC JTC1/SC27 is concerned with "Security Techniques" and reporting to this are three working groups, of which ISO/IEC JTC1/SC27/WG3 is concerned with "Security Evaluation Criteria". It has taken over a year to reach this point and a further eight years at least will be needed to produce a definitive standard. This particular strand of the efforts in the field looks to be too slow to have much effect but remember the tortoise and the hare.

By contrast, the two European efforts are moving very quickly. The ITSEC appeared in May 1990 and was publicly reviewed in October of the same year. Version 1.1 appeared in January 1991 and was reviewed on April 17th. Version 1.2 was issued in August 1991 and is not expected to change significantly for two or three years. During those two or three years the Commission of the European Communities via its RACE program will be producing a set of criteria which will, it is to be hoped, subsume the ITSEC and become the European standard. If this is successful then it could well become the *de facto* world commercial computer security standard because of the policy problems with the Orange Book [TCSEC] and its derivatives.

A *de facto* standard could, of course, become a *de jure* ISO standard through the accelerated procedure in less than three years which would total six years from now and be quicker than the JTC1 route!

For those of us with an interest in this field it is certainly very interesting and exciting to have three or four competing efforts which may well cross-fertilise to produce a good workable standard within the forseeable future.

References

[BandP] *David Bell and L. J. La Padula:* Secure Computer System: Unified Exposition and MULTICS Interpretation. USA 1975, Mitre Report ESD-TR-75-306.

[BIBA] *K. J. Biba:* Integrity Considerations for Secure Computer Systems. USA 1977, Mitre Report ESD-TR-76-372.

[CandW] *D. D. Clark and D. R. Wilson:* A Comparison of Military and Commercial Security Policies. Proceedings of the 1987 IEEE Symposium on Security and Privacy.

[CESG3] *Communication-Electronics Security Group:* UK Systems Security Confidence Levels. CESG Memorandum No. 3. United Kingdom, February 1989.

[DTIEC] *Department of Trade and Industry Commercial Computer Security Centre:* Overview of Documentation (V01). Great Britain, February 1989 (DRAFT only).
Glossary (V02). Great Britain, February 1989 (DRAFT only).
Users' Code of Practice (V11). Great Britain, November 1989 (DRAFT only).
Security Functionality Manual (V21). Great Britain, February 1989 (DRAFT only).
Evaluation Levels Manual (V22). Great Britain, February 1989 (DRAFT only).
Evaluation and Certification Manual (V23). Great Britain, February 1989 (DRAFT only).
Vendors' Code of Practice (V31). Great Britain, November 1989 (DRAFT only).
Colloquially known as *the Green Books.*

[FREC] *Service Central de la Sécurité des Systèmes d'Information:* Critères Destinés à Évaluer le Degré de Confiance des Systèmes d'Information. 692/SGDN/DISSI/SCSSI. France, July 1989.

[ITSEC] *Der Bundesminister des Innern:* Information Technology Security Evaluation Criteria (Harmonised Criteria of France - Germany - the Netherlands - the United Kingdom). Federal Republic of Germany, Version 1.1, January 1991.

[ROBER] *D. W. Roberts:* Computer Security: Policy, Planning and Practice. ISBN 0-86353-180-6. NCC Blackwell, United kingdom, January 1990.

[TCSEC] *Department of Defense:* Trusted Computer Systems Evaluation Criteria. DOD 5200.28-STD. United States of America, December 1985. Colloquially known as *the Orange Book.*

[ZSIEC] *Zentralstelle für Sicherheit in der Informationstechnik:* Criteria for the Evaluation of Trustworthiness of Information Technology (IT) Systems. ISBN 3-88784-200-6. Federal Republic of Germany, January 1989. Also sometimes colloquially known as *the Green Book.*

Abbreviations, Nicknames and Acronyms

ADP	Automatic Data PROCESSING
BSI	British Standards Institute
CCSC	DTI Commercial Computer Security Centre (UK Government)
CESG	Communications-Electronics Security Group of GCHQ (UK Government)
DoD	Department of Defense (US Government)
DTI	Department of Trade and Industry (UK Government)
GCHQ	Government Communications Headquarters (UK Government)
Green Book	Often used by Germans (and sometimes used by others) as the nick-name of the German Criteria [ZSIEC]
Green Books	Confusingly, often used in Europe as the nick-name of the UK DTI Criteria [DTIEC]
HCI	Human-Computer Interface
IEC	International Electrotechnical Commission
IEEE	Institute of Electrical and Electronic Engineering
ISBN	International Standard Book Number
ISO	International Standards Organisation
IT	Information Technology
MoD	Ministry of Defence (UK Government)
NBS	US National Bureau of Standards (now known as NIST)
NIST	US National Institute for Science and Technology (formerly NBS)
Orange Book	Always used to denote the US DoD Criteria [TCSEC] (Except when used for the CCITT Orange Books)
UK	United Kingdom of Great Britain and Northern Ireland
US USA	United States of America United States of America

Standardization of Cryptographic Techniques

Bart Preneel*

ESAT-COSIC Laboratory, K.U.Leuven
K. Mercierlaan 94, B-3001 Leuven, Belgium

bart.preneel@esat.kuleuven.ac.be

Abstract. An overview of standardization activities in the field of cryptography is given, including a description of worldwide, European, and North-American standard organizations. More details are given on the status of the work on open systems and within the committees ISO/TC68 and ISO/IEC JTC1/SC27.

1 Introduction

The evolution of cryptography and computer security from military and diplomatic to commercial applications has created a demand for standardization in this area. The goal of this standardization is to offer secure solutions, to increase the competition and to achieve the economy of scale. Several levels of standards can be distinguished [50]:

base standards: for generic user requirements,
functional standards: for procurement, product certification, and services; they explain the common industry approach to use base standards,
evaluation criteria: for the evaluation of products and systems,
industry standards and codes of practice: technical and procedural standards that are required by specific user groups or industries,
interpretative documents: guidelines, glossaries for information and education.

A large number of areas of standardization in security has been identified in [50]. The standards that will be discussed here are mainly base standards on cryptographic techniques; they can be divided into several subareas:

cryptographic algorithms: it is essential that the algorithms are public. This contrasts to the old approach in cryptography, where the algorithms were kept secret. The advantage of this approach is that the algorithms can be thoroughly evaluated within the international research community, but the other side of the medal is that they offer an interesting target for the organized crime. The standardization of algorithms protecting confidentiality

---

* NFWO aspirant navorser, sponsored by the National Fund for Scientific Research (Belgium).

can cause problems because certain countries believe this conflicts with their national interests. This has resulted in a registration of algorithms for confidentiality protection (cf. Sect. 5).

modes of use: this specifies how to use cryptographic algorithms in order to achieve confidentiality, message authentication, entity authentication, ... A second element is the use of cryptographic algorithms for key management. These standards are partially algorithm independent.

applications: within specific applications like banking and telecommunications a detailed specification has to be made, resulting in a completely different type of standards. They include the selection of algorithms and parameters, but also the specification of all interfaces.

This paper only intends to give an idea of the nature of the work, and of some important developments. It does not have the ambition to be complete. First organizations involved in security standardization will be described. Subsequently the status of security standardization of open systems and of security techniques are briefly reviewed and our conclusions are presented.

2 Standardization Organizations

In this section a very brief overview will be given of organizations that are involved in the standardization of security techniques and of organizations that use security techniques as part of their standards. For a more detailed treatment the reader is referred to [50].

2.1 Worldwide Organizations

The three important worldwide standardization organizations are ISO (International Standards Organization), IEC (International Electrotechnical Commission), and CCITT (International Telegraphy and Telephone Consultative Committee). This list is not complete: e.g., standards for EDI (Electronic Date Interchange) are developed under the auspices of the United Nations/Economic and Social Council (UN/EDIFACT Security Joint Working Group).

ISO/IEC. The member bodies of ISO are the national standardization organizations of the 74 participating countries. Within ISO, the work is carried out in Technical Committees, that establish Subcommittees (SC). The work in a SC is allocated to several Working Groups (WG). Special working groups (SWG) are on the same level as SC's. Before a document is published as an International Standard (IS), it goes through several stages: Study Period, New work item Proposal (NP), Working Draft (WD), Committee Draft (CD), Draft International Standard (DIS), and finally International Standard (IS). Under the most optimal circumstances, an international standard can be developed in three years. In most cases, five years is a more realistic estimate. ISO is a hierarchical organization; at every level the member bodies take the decisions (e.g., vote on

NP, CD, DIS). They also send experts who perform the technical work in the Working Groups.

ISO and IEC cooperate within JTC1, a joint technical committee on "Information Technology". Activities in JTC1 are divided over several groups, subcommittees (SC) and working groups (WG). The groups and committees that perform security related standardization activities are the Special Working Groups on EDI and on Security, SC6 (OSI Lower Layers), SC17 (Identification Cards), SC18 (Text and Office Systems), SC21 (OSI Architecture, Management, and Upper Layers), SC22 (Languages), and SC27 (Security Techniques).

A separate ISO Committee exists with the title "Banking and Related Financial Services". Security related activities are carried out in SC2 (Operations and Procedures) within WG2 (Security for Wholesale Banking) and in SC6 (Financial Transaction Cards, Related Media, and Operations) within WG6 (Security in Retail Banking) and WG7 (Security Architecture of Banking Systems using the Integrated Circuit Card).

CCITT. CCITT is concerned with data communications issues, including the development of standards. CCITT is a part of the International Telecommunication Union (ITU), a United Nations organization. The standards produced by CCITT are called recommendations and are updated every 4 years. Technical activities are allocated over 17 study groups. Security related work is taking place in SG VII (Data Networks).

2.2 Regional Organizations

Europe.

CEN/CENELEC: European Committee for Standardization, corresponding to ISO and IEC respectively.

ETSI: European Telecommunications Standards Institute. Topics include Smart Cards, X.400 Security, Mobile services, Network Management, and Audio Visual Services.

ITAEGV: Information Technology Advisory Expert Group for Information Security. This is a group of experts of CEN, CENELEC, and ETSI that works on the specification of security requirements for future standardization work; this group has produced a "Taxonomy for Standardization of Security" [50].

EWOS: European Workshop for Open Systems. Work on X.400 MHS security profile.

ECMA: European Computer Manufactures Association. Work on security of database systems, OSI lower layer security, Distributed Office Applications (DOA), Open system security. Within the EEC RACE project SESAME (Secure European System for Applications in a Multivendor Environment) an architecture is developed that implements the ECMA model [60].

North America.

ANSI: American National Standards Institute. Work on EDI, banking, techniques/mechanisms, and OSI security. The standards in the field of Information Systems have a number starting with an X.

IEEE: Institute of Electrical and Electronic Engineers. Work on LAN (Local Area Network) security [28] and POSIX (Portable Operating System Interface for Computer Environments) security.

NIST: National Institute for Standards and Technology (formerly the NBS, National Bureau of Standards). NIST has issued the Federal Information Processing Standards, including the famous Data Encryption Standard (DES) [15] and related standards [16, 17]. Standards have also been published on password usage [18], message authentication [19], and integrity guidelines. Other standards concerning the use of DES were published by the General Services Administration (GSA) [22, 23].

Important recent developments are the publication of two draft standards, namely the Digital Signature Standard [20], specifying the Digital Signature Algorithm or DSA, and the Secure Hash Standard [21] specifying the Secure Hash Algorithm [21]. Note that for these standards a distinction is made between the standard and the algorithm specified in the standard. For more details on these drafts the reader is referred to [62]. It is very remarkable that RSA, which has become an industrial de facto standard, has not been selected by NIST. The publication of DSA has led to a controversial discussion with technical, economical, and political arguments. Note that the draft digital signature under development within ANSI committee X9 provides the choice between both algorithms.

NIST/OIW: NIST OSI Implementors Workshop.

2.3 Other Organizations

To this category belong the US/Dod (Department of Defense) and the US National Computer Security Center (NCSC) that have published standards on evaluation criteria: Trusted Computer System Evaluation Criteria (TCSEC), better known as the Orange Book [14], Trusted Network Interpretation (TNI, Red Book [57]), and Trusted Database Interpretation (TDI, Gray Book, classified). Four European countries (United Kingdom, France, Germany and The Netherlands) are currently developing their Information Technology Security Evaluation Criteria (ITSEC) [51]. The corresponding ITSEM draft [52] specifies the evaluation procedures. More details can be found in [63].

A second class are the Government Procurement Specifications, to help the government departments specify and purchase open systems equipment: GOSIP (Government OSI Profile) for UK, USA, and Canada, EPHOS (European Procurement Handbook for Open Systems), a European GOSIP.

Other organizations that produce standards are the US National Security Agency (NSA) together with NIST who issued the SDNS (Secure Data Network Systems) for OSI security [58, 59], the NATO that has produced NOSA (NATO

Security Architecture), and S.W.I.F.T. (Society for Worldwide International Financial Telecommunications).

Influence on standardization work can also be expected from collaborative research: RACE (Research & Development in Advanced Communications in Europe), ESPRIT (European Strategic Programme for Research & Development in Information Technologies), TEDIS (Trading EDI System), AIM (Advanced Informatics in Medicine), and ANSA/ISA (Advanced Network Systems Architecture). Other contributions originate within user groups and vendor associations.

3 Open Systems

The development and use of open systems standards can be defined as *"The harmonisation of communication, distributed systems and processing standards to allow interworking independent of the nature of the systems involved"* [50]. A system is said to be open if it can interwork with another system using open system standards. The standardization of open systems is being carried out under the headings Architecture, Frameworks, and Models; Service and Protocol Extensions; Techniques and Mechanisms; Applications; Management. The subcommittees involved are ISO/IEC JTC1/SC21, SC6, and SC27.

Architecture: The OSI Security Architecture ISO/IEC 7498-2 [29] is one of the fundamental standards for open systems interconnection. This standard provides the base communications architecture for the development of open systems and it was given its international standard status in February 1988. This standard is also being adopted as a CCITT recommendation as part of their commitment to open systems.

Frameworks: Frameworks provide a generic solution to a specific security problem and can be used wherever that problem exists; or more specifically, the frameworks can be used to ensure consistency in the security enhancements to protocols. These frameworks include: authentication, access control, integrity, non-repudiation, confidentiality, and audit. They are still under development.

Models: The purpose of the OSI models is to apply the security concepts detailed in the OSI Architecture and the Security Frameworks to two specific ares of OSI: the upper layers and the lower layers.

Protocols and Service Standards: Security enhancements to protocols and service standards are being developed. Only for the physical layer a standard has been published (IS 9160 [35]).

Security Techniques: Work on security techniques comprises authentication (entity authentication and data origin), data integrity, digital signatures, hash functions, non-repudiation, and key management.

Applications: The most important standards dealing with security in distributed applications are X.400 [8] and X.509 [10]. X.400 is a Message Handling System that enables users to exchange messages on a store and forward basis. In the 1988 CCITT X.400 recommendations, the following security services have been specified: authentication, integrity, non-repudiation, and confidentiality. Among the distributed applications that can be supported are email and EDI. An almost equivalent standard has been adopted within JTC1/SC18: MOTIS. The famous X.509, the authentication framework for Directory Systems, complements the security features defined by X.400. This framework provides protocol definitions for both weak authentication mechanisms based on passwords and strong authentication mechanisms based on the use of asymmetric encryption techniques. It has been adopted as an ISO-standard (ISO/IEC 9594-8 [36]). Note that the hash function proposed in annex D is insecure [11]. The authentication mechanisms have been criticized in [27, 56]. CCITT are currently working on recommendations for EDI (X.435) that will employ X.400 security and additional security measures.

Management: The security related aspects to the management of open systems are the management of security and the management use of security.

4 ISO/TC68

The ISO Technical Committee 68 on "Banking and Related Financial Services" has produced standards on message authentication and key management. The standards for authentication for wholesale banking contain an general model [31] and specify three Message Authentication Codes (MAC's) [32, 33]. The first two are the CFB and the CBC mode of DES, similar to ANSI X9.9 [3], and the third is the Message Authentication Algorithm. A standard for key management for wholesale banking has also been produced [34], based on ANSI X9.17 [4]. Other standards specify procedures for message encipherment (IS 10126), retail message authentication (DIS 9807 [42]), and PIN management (IS 9564). The relevant standards under development include standards for retail key management (CD 11568), and key management by means of asymmetric algorithms (CD 11166). More details on the current status of this draft can be found in [64]. For an overview of standardization in the area of smart cards, the reader is referred to [26].

5 ISO/IEC JTC1/SC27

Recently two important changes have been made to the organization of the work on security within the ISO world. First, in 1987 the work within ISO/TC97 "Information Processing" was moved to a new committee JTC1 with the title "Information Technology". Within JTC1, SC27 "Security Techniques" has replaced in 1989 the subcommittee SC20 "Data Cryptographic Techniques"; at the same time its scope was widened.

It is important to remark that the scope of SC27 is not restricted to security for open systems. The work in SC27 is currently allocated to three working groups:

WG1 Requirements, Security Services and Guidelines

WG2 Techniques and Mechanisms
WG3 Security Evaluation Criteria

WG1 and WG3 had their first meeting in October 1990, while WG2 is more or less the continuation of both working groups of the old SC20.

WG3 is planning to develop standards for IT security evaluation and certification of IT systems, components, and products. This comprises evaluation criteria, and a methodology for application of the criteria and administrative procedures for the evaluation, certification, and accreditation schemes. Current projects are "Evaluation criteria for IT security" and "Collection and analysis of requirements for IT security evaluation criteria".

WG1 has as tasks to identify application and system requirement components (e.g., risk analysis), to develop standards for security services, and to develop the supporting interpretative and guiding documents. Current projects are "Security information objects", "Guidelines for the management of IT security", "Glossary of IT security terminology", and "Key management framework",

WG1 is also maintaining the standard that describes the procedures for the registration of cryptographic algorithms [43]. For the time being several algorithms have already been registered. It must be stressed that appearance of an algorithm on the register implies no guarantee of level of security.

WG2 has taken over most of the work of WG1 and WG2 of the former SC20, but its scope has been extended to non-cryptographic techniques (e.g., access control, passwords). The work items that are currently under the responsibility of SC27/WG2 are:

Modes of Operation: The four most common modes of operation (ECB, CBC, CFB, and OFB) have been standardized for a 64-bit algorithm [30]. A generalization for an n-bit algorithm was also published [44]. At the same time the presentation was improved: it contains annexes describing the properties of the different modes, specifies optional extensions of the CBC mode, and gives some examples [44].

Data Integrity: This standard specifies a Message Authentication Code or MAC; its full title is "Data integrity mechanism using a cryptographic check function employing an n-bit algorithm with truncation" [38]. It is based on ANSI X9.19 [5] and ISO 8731-1 [32]. The main differences are that only the CBC mode is selected, and that it can be applied to any block cipher. Because of a defect report (the description of the padding scheme was ambiguous), a new version is under development that contains some other improvements. This version is currently at the DIS stage.

Entity authentication: Several parts of a standard for entity authentication have already been completed [39, 40, 41]. Part 1 describes the general model and has become an IS. Part 2 discusses entity authentication using symmetric techniques and is in the DIS stage, while part 3 dealing with entity authentication using public techniques has been published recently as an IS. New parts on using non-reversible functions and based on zero-knowledge techniques are under development.

Digital Signatures: One standard specifying a digital signature giving message recovery has been published [37]. It defines the redundancy that has to be introduced to destroy the homomorphic properties of RSA and its extension to even exponents [25]. This standard is only suited for signing short messages. A second standard will specify digital signature schemes that can be applied to messages of arbitrary size, i.e., digital signature schemes that are combined with a hash function.

Hash Functions: Both part 1 (general) and part 2 (hash-function using an n-bit block cipher algorithm) are in the DIS stage [45, 46]. The hash functions in part 2 are the scheme of S. Matyas, C. Meyer, and J. Oseas [54] and MDC-2 [7], also known as the Meyer-Shilling hash function (for more details the reader is referred to [62]). Hash functions using modular arithmetic and dedicated hash functions are only in the NP stage.

Non-repudiation: A three part standard is under development; it is currently in the working draft phase. Part 1 will specify the general model, part 2 will specify non-repudiation using symmetric techniques, and part 3 will specify non-repudiation using asymmetric techniques.

Zero Knowledge Techniques: It was recently decided to incorporate the current working draft documents on zero knowledge into the work on entity authentication and digital signatures.

Key Management: A multi-part standard is under development [47, 48, 49]. Part 1, describing the key management framework will be developed within WG1. At the same time it should contain the common elements of parts 2 and 3, which might lead to problems. Part 2 will specify key management mechanisms using symmetric techniques and is in the CD stage. The main problems for the moment are the compatibility with part 2 of entity authentication [40] and with the existing banking standards (ISO 8732 [34] and ANSIX9.17 [4]) that combine key establishment with key notarization. Moreover some countries request that the amount of encrypted material is minimized. Part 3 specifying key management mechanisms using asymmetric techniques is expected to enter the CD stage soon. In this case the corresponding banking standards are still under development.

6   Conclusion

International standardization of security techniques is currently concentrated in the committee ISO/IEC JTC1/SC27 and ISO/TC68 (only for financial applications). The recent restructuring of the work has certainly increased the efficiency, but developing an international standard is a slow process: the basic idea is to build a consensus, which is not facilitated by the fact that national bodies can change their position at any moment in time. A variety of organizations is standardizing the use of security techniques for providing secure computer systems, secure networks, and secure applications.

References

1. ANSI X3.92-1981, *"American National Standard for Information Systems – Data Encryption Algorithm (DEA),"* ANSI, New York.

2. ANSI X3.106-1983, *"American National Standard for Information Systems — Modes of Operation,"* ANSI, New York.

3. ANSI X9.9-1986 (Revised), *"American National Standard for Financial Institution Message Authentication (Wholesale),"* ANSI, New York.

4. ANSI X9.17-1985, *"American National Standard for Financial Institution Key Management (Wholesale),"* ANSI, New York.

5. ANSI X9.19-1986, *"American National Standard for Financial Institution Key Management (Retail),"* ANSI, New York.

6. ANSI X9.30, *"Digital Signatures for Financial Data Using Public Key Cryptography,"* ANSI, New York, April 20, 1991.

7. B.O. Brachtl, D. Coppersmith, M.M. Hyden, S.M. Matyas, C.H. Meyer, J. Oseas, S. Pilpel, and M. Schilling, *"Data Authentication Using Modification Detection Codes Based on a Public One Way Encryption Function,"* U.S. Patent Number 4,908,861, March 13, 1990.

8. C.C.I.T.T. X.400, *"Message Handling/Information Processing Systems,"* C.C.I.T.T. Recommendation, 1988.

9. C.C.I.T.T. X.500, *"The Directory – Overview of Concepts,"* C.C.I.T.T. Recommendation, 1988, (same as ISO/IEC 9594-1, 1989).

10. C.C.I.T.T. X.509, *"The Directory – Authentication Framework,"* C.C.I.T.T. Recommendation, 1988, (same as ISO/IEC 9594-8, 1989).

11. D. Coppersmith, "Analysis of ISO/CCITT Document X.509 Annex D," *IBM T.J. Watson Center, Yorktown Heights, N.Y., 10598, Internal Memo,* June 11, 1989, (also ISO/IEC JTC1/SC20/WG2/N160).

12. D. Davies and W.L. Price, *"Security for Computer Networks: an Introduction to Data Security in Teleprocessing and Electronic Funds Transfer (2nd edition),"* Wiley & Sons, 1989.

13. M. De Soete and K. Vedder, "Authentication standards," *Proc. of the 3rd symposium on State and Progress of Research in Cryptography,* W. Wolfowicz, Ed., Fondazione Ugo Bordoni, 1993, pp. 207–218.

14. DoD 5200.28-STD, *"Trusted Computer Systems Evaluation Criteria,"* Department of Defense, U.S.A., December 1985.

15. FIPS 46, *"Data Encryption Standard,"* Federal Information Processing Standard, National Bureau of Standards, U.S. Department of Commerce, Washington D.C., January 1977.

16. FIPS 74, *"Guidelines for Implementing and Using the NBS Data Encryption Standard,"* Federal Information Processing Standard, National Bureau of Standards, US Department of Commerce, Washington D.C., April 1981.

17. FIPS 81, *"DES Modes of Operation,"* Federal Information Processing Standard, National Bureau of Standards, US Department of Commerce, Washington D.C., December 1980.

18. FIPS 112, *"Password Usage,"* Federal Information Processing Standards, National Bureau of Standards, National Bureau of Standards, US Department of Commerce, Washington D.C., May 1985.

19. FIPS 113, *"Computer Data Authentication,"* Federal Information Processing Standard, National Bureau of Standards, US Department of Commerce, Washington D.C., May 1985.

20. FIPS xxx, *"Digital Signature Standard,"* Federal Information Processing Standard, Draft, National Institute of Standards and Technology, US Department of Commerce, Washington D.C., August 30, 1991.

21. FIPS yyy, *"Secure Hash Standard,"* Federal Information Processing Standard, Draft, National Institute of Standards and Technology, US Department of Commerce, Washington D.C., January 31, 1992.

22. FS 1027, *"Telecommunications: General security requirements for equipment using the Data Encryption Standard,"* Federal Standard, General Services Administration, April 1982.

23. FS 1028, *"Interoperability and Security Requirements For Use of the Data Encryption Standard With CCITT Group 3 Facsimile Equipment,"* Federal Standard, General Services Administration, April 1985.

24. L.C. Guillou, M. Davio, and J.-J. Quisquater, "Public-Key Techniques: Randomness and Redundancy," *Cryptologia*, Vol. 13, April 1989, pp. 167-189.

25. L.C. Guillou, J.-J. Quisquater, M. Walker, P. Landrock, and C. Shaer, "Precautions taken against various potential attacks in ISO/IEC/DIS9796," *Advances in Cryptology, Proc. Eurocrypt'90, LNCS 473*, I.B. Damgård, Ed., Springer-Verlag, 1991, pp. 465–473.

26. L.C. Guillou, M. Ugon, and J.-J. Quisquater, "The smart card: a standardized security device dedicated to public cryptology," in *"Contemporary Cryptology: The Science of Information Integrity,"* G.J. Simmons, Ed., IEEE Press, 1991, pp. 561–613.

27. C.I'Anson and C.J. Mitchell, "Security defects in CCITT recommendation X.509 – The Directory – Authentication Framework," *ACM Computer Communication Review*, Vol. 20, No. 2, 1990, pp. 30–34.

28. IEEE 802.10, *"Standards for Interoperable Local Area Network (LAN) Security (SILS),"* Unapproved Draft IEEE Standard P802.10, September 12, 1989.

29. ISO/IEC 7498-2, *"Information processing – Open systems interconnection – Basic reference model – Part 2: Security architecture,"* ISO/IEC, 1987.

30. ISO 8372, *"Information processing - Data cryptographic techniques - Modes of operation for a 64-bit block cipher algorithm,"* ISO, 1987.

31. ISO 8730, *"Banking - Requirements for message authentication (wholesale),"* ISO, 1990.

32. ISO 8731-1, *"Banking - approved algorithms for message authentication - Part 1: DEA,"* ISO, 1987.

33. ISO 8731-2, *"Banking - approved algorithms for message authentication - Part 2: Message Authentication Algorithm (MAA),"* ISO, 1987.

34. ISO 8732, *"Banking - Key management (wholesale),"* ISO, 1989.

35. ISO/IEC 9160, *"Information Technology - Data cryptographic techniques - Physical layer interoperability requirements,"* ISO/IEC, 1988.

36. ISO/IEC 9594-8, *"Information technology - Open systems interconnection - The directory - Part 8: Authentication framework,"* ISO/IEC, 1990.

37. ISO/IEC 9796, *"Information technology - Security techniques - Digital signature scheme giving message recovery,"* ISO/IEC, 1991.

38. ISO/IEC 9797, *"Information technology - Data cryptographic techniques - Data integrity mechanisms using a cryptographic check function employing a block cipher algorithm,"* ISO/IEC, 1989.

39. ISO/IEC 9798-2, *"Information technology - Security techniques - Entity authentication - Part 1: General model,"* ISO/IEC, 1991.

40. ISO/IEC DIS 9798-2, *"Information technology - Security techniques - Entity authentication - Part 2: Entity authentication using symmetric techniques,"* ISO/IEC, 1993.

41. ISO/IEC 9798-3, *"Information technology - Security techniques - Entity authentication - Part 3: Entity authentication using a public key algorithm,"* ISO/IEC, 1993.

42. ISO DIS 9807, *"Banking and related financial services - Requirements for message authentication (retail),"* ISO, 1992.

43. ISO/IEC 9979, *"Information technology - Security techniques - Procedures for the registration of cryptographic algorithms,"* ISO/IEC, 1991.

44. ISO/IEC 10116, *"Information technology - Security techniques - Modes of operation of an n-bit block cipher algorithm,"* ISO/IEC, 1991.

45. ISO/IEC DIS 10118-1, *"Information technology - Security techniques - Hash-functions - Part 1: General,"* ISO/IEC, 1992.

46. ISO/IEC DIS 10118-2, *"Information technology - Security techniques - Hash-functions - Part 2: Hash-functions using an n-bit block cipher algorithm,"* ISO/IEC, 1992.

47. ISO/IEC WD 11770, *"Information technology - Security techniques - Key management - Part 1: Framework,"* ISO/IEC SC27/N644, 1992.

48. ISO/IEC CD 11770, *"Information technology - Security techniques - Key management - Part 2: Key management mechanisms using symmetric techniques,"* ISO/IEC, 1992.

49. ISO/IEC WD 11770, *"Information technology - Security techniques - Key management - Part 3: Key management mechanisms using asymmetric techniques,"* ISO/IEC SC27/WG2/N190, 1993.

50. ITAEGV, *"Taxonomy of security standardisation,"* Version 2.0, ITAEGV/N69, April 1992.

51. ITSEC, *"Information Technology Security Evaluation Criteria (ITSEC),"* Provisional Harmonised Criteria, Version 1.2, June 1991.

52. ITSEM, *"Information Technology Security Evaluation Manual (ITSEM),"* Version 0.2, April 1992.

53. J. Lindballe, "Standardization and Data Security," *Proceedings Secubank '88*, Datakontext-Verlag, pp. 286–299.

54. S.M. Matyas, C.H. Meyer, and J. Oseas, "Generating strong one-way functions with cryptographic algorithm," *IBM Techn. Disclosure Bull.*, Vol. 27, No. 10A, 1985, pp. 5658–5659.

55. C.H. Meyer and M. Schilling, "Secure program load with Manipulation Detection Code," *Proc. Securicom 1988*, pp. 111–130.

56. C. Mitchell, M. Walker, and D. Rush, "CCITT/ISO standards for secure message handling," *IEEE Journal on Selected Areas in Communications*, Vol. 7, May 1989, pp. 517–524.

57. NCSC TNI, *"Trusted Network Interpretation,"* National Computer Security Center, July 1987.

58. SDN.301, *"Secure Data Network Systems Security Protocol 3 (SP3),"* National Security Agency and National Institute of Standards and Technology, Revision 1.5, May 1989.

59. SDN.401, *"Secure Data Network Systems Security Protocol 4 (SP4),"* National Security Agency and National Institute of Standards and Technology, Revision 1.3, May 1989.

60. D. Pinkas, T. Parker, and P. Kaijser, "SESAME (Secure European System for Applications in a Multivendor Environment): an Introduction", *Technical Report*, February 1993.

61. W.L. Price, "Progress in data security standardization," *Advances in Cryptology, Proc. Crypto'89, LNCS 435*, G. Brassard, Ed., Springer-Verlag, 1990, pp. 620–623.

62. B. Preneel, R. Govaerts, and J. Vandewalle, "Information authentication: hash functions and digital signatures," *this volume.*

63. D.W. Roberts, "Evaluation criteria for IT security," *this volume.*

64. R.A. Rueppel, "Criticism of ISO CD 11166, Banking - Key management by means of asymmetric algorithms," *Proc. of the 3rd symposium on State and Progress of Research in Cryptography*, W. Wolfowicz, Ed., Fondazione Ugo Bordoni, 1993, pp. 191–198.

65. M.E. Smid and D.K. Branstad, "The Data Encryption Standard: Past and Future," in *"Contemporary Cryptology: The Science of Information Integrity,"* G.J. Simmons, Ed., IEEE Press, 1991, pp. 45–64.

Numbers Can Be a Better Form of Cash Than Paper

David Chaum

Centre for Mathematics and Computer Science
Kruislaan 413, NL-1098 SJ Amsterdam, The Netherlands

Soon, by accessing a computerized network from almost anywhere, you may be able to pay for a purchase, change your insurance coverage, or perhaps even send an electronic "letter" to a friend. Although a single system integrating all these functions is still some way off, its piecemeal construction is already underway. Automatic cash dispensers and electronic payment devices at shops, for instance, are in use in many countries and are planned in many more. The technology underlying all this does have enormous potential for cutting organizations' costs and increasing their security, as well as for enhancing consumer convenience. The prevailing approach to applying the technology, however, brings with it some quite serious dangers.

This current approach requires individuals to identify themselves to the system each time they use it. All the various identifying techniques—like tamper-resistant plastic cards, memorized secret numbers, and fingerprints—are essentially equivalent to universal ID numbers, such as the ones used for social insurance or passports. These identifiers allow computerized linking of all manner of personal information, from school, medical, and employment records to purchase details captured in electronic payments. Faced with how easily such data can be tapped, exchanged, and modified, legal mechanisms seem powerless to protect individuals from errors and data misuse. Moreover, "automatic pattern recognition" techniques could be applied by anyone tapping into the large-scale systems currently being planned. Individuals might thus be categorized by their transaction patterns—everywhere they pay, every relationship they have with organizations, and everyone they communicate with—in a form of invisible mass surveillance.

All these problems can be avoided by a new approach to using so-called smart cards—plastic cards containing microcomputer chips. The most technically advanced of these, the "supersmart" card made by Toshiba for Visa, is no bigger or thicker than the familiar credit card, yet includes a battery, character display, and buttons like a pocket calculator. As with all current- approach cards, though, the organization that owns and issues it must make it tamper-resistant to prevent anyone else from accessing its internal structure. By contrast, the new-approach card computer is all yours. You can choose one just as you would any pocket-sized calculator or personal computer, and can even access or customize its inner workings to suit your convenience. Tamper resistance, with its inflexibility, expense, and low security, has been made unnecessary by more advanced coding techniques.

1 Shopping With Your Electronic Wallet

When you buy something with your new-approach card computer, the clerk's electronic cash register transmits to your card the cost and description of the purchase. If you agree to these details as displayed by your card, all you do is enter on its keyboard your single secret authorizing number. Your card then completes the transaction by transmitting to the cash register a one-time-use number of several hundred digits–a number that is money.

Such a card not only improves on the personal convenience and security of credit cards, but the coding it uses ensures privacy. Even paper cash is traceable in principle, since the bank could record the serial numbers of the notes you withdraw. Conceptually, the inherent traceability of pre-printed notes could be avoided by instead, during withdrawal, having the bank validate and return envelopes that you supply. A plain slip within such an envelope would get, say, a carbon-paper image of the bank's "worth-one-dollar" validating signature stamp. You could then discard the envelope and spend your validated yet untraceable slip just like cash. The actual digital system works essentially the same way: your card randomly codes a number that it chooses to serve like the slip, obtains the bank's coded validating signature on it during a withdrawal, and removes the random coding before spending the resulting validated numeric note. A simple mathematical proof shows that the bank can't trace such a numeric note to its withdrawal, no matter how extensive or ingenious the computerized analysis. If you need to, though, you can reveal information that lets any one of your payments be traced incontestably.

Security for banks and merchants is also improved. The bank's coded signature, which lets a numeric note's validity be tested by any card or cash register, is far harder to counterfeit than printed money. But because numbers are easy to copy exactly, retailers need protection against someone spending a note number more than once. For high-value payments, the bank's list of already spent notes is electronically consulted while you wait. For low-value transactions, shops avoid this expense by requiring a random selection of additional numeric information from the payer. The amount of such information revealed in spending a note once is absolutely useless in tracing the payer. But enough additional information is revealed in spending a note twice so that, after the day-end deposit of notes, the bank gets from each "double spender" the equivalent of a signed confession.

2 Showing Credentials Without Identification

Business and government organizations sometimes do legitimately need to see statements that other organizations have issued about individuals. Such credentials have in the past taken the form of identifying paper certificates, like passports, driver's licenses, and membership cards. The original purpose of most of these documents was to securely authenticate individuals' qualifications, such as an economic or age bracket, a license, or an academic degree. Identification was a means to that end. But in today's computerized imitation of paper-based methods, identification allows organizations to match with or directly access other organizations' computerized files. This has created in effect a single huge database on individuals, although it remains slow, incomplete, and prone to error. Yet a comprehensive and centralized system, even if it were acceptable to the public, would be prohibitively expensive.

An extension of the card-computer payment technique can solve these problems. All credential transactions are conducted through your card, using a different numeric alias or "digital pseudonym" with each organization. Credentials are issued in the form of unforgeable coded signatures on these pseudonyms, which your card handles much as it does numeric notes. If you authorize it to, your card can transmit convincing numeric proof that you hold at least one combination of credential signatures meeting an organization's requirement–without revealing anything more. The way the pseudonyms are created assures organizations that you can't lend, modify, or escape accountability for your credentials. And because of the way your card codes "blank" credentials in "envelopes" before they are signed by organizations, your pseudonyms cannot be linked any more than your payments can. You retain complete control over your personal information, just as if all the computerized records that organizations maintain on you today were stored only in your card computer.

3   Protecting Electronic Mail

Once electronic payment becomes commonplace, electronic mail may be next. The widespread use of "e-mail" may raise a number of problems: keeping message content confidential, preventing users from falsely denying that they have sent or received particular messages, and forestalling automated "traffic analysis" that could trace the patterns of users' relationships.

The new approach's solution broadcasts messages to all electronic mail computers in the network. This prevents tracing to the intended recipient, and it ensures that receipt of messages cannot be denied. Also, a novel kind of coding permits each person's e-mail computer to conceal unconditionally which messages it sends. Other coding keeps message content confidential and lets recipients be sure–and convince others if necessary–of the message originator's digital pseudonym. This solution naturally facilitates payment and credential transactions from home, with all the convenience and protection offered by card computers.

4   The Moment of Decision

Lower cost and higher security will serve as strong motivation for organizations to adopt the new approach. As more people become aware both of the dangers posed by the current approach and of the improved personal convenience and security offered by the alternative, organizations may be further motivated by consumer preference, public opinion, and even legislation.

At the moment, however, the centralized-data approach is gaining momentum. Automatic cash dispensers and other current electronic payment techniques, with the tracing they allow, are just the tip of the massive investment iceberg required by such large-scale systems.

We are fast approaching a moment of crucial and perhaps irreversible decision, not merely between two kinds of technological system, but between two kinds of society. Current developments in applying technology are rendering hollow both the remaining safeguards on privacy and the right to access and correct personal data. If these developments continue, their enormous surveillance potential will leave individuals' lives vulnerable to an unprecedented concentration of scrutiny and authority. If, on the other hand, the new approach prevails, the erosion of our informational rights can be reversed and new rights added–notably the right, realizeable through personal card computers, to reveal only necessary information in transactions. As we move into an age of pervasive computerization, control over information becomes the key to social, economic, and political power. Card computers can restore balance by putting part of that key, both literally and figuratively, back in the hands of private citizens.

References

1. D. Chaum, "Blind Signatures for Untraceable Payments," *Advances in Cryptology, Proc. Crypto'82*, D. Chaum, R.L. Rivest, and A.T. Sherman, Eds., Plenum Press, New York, 1983, pp. 199–203.
2. D. Chaum, "Security Without Identification: Transaction Systems to Make Big Brother Obsolete," *Communications of the ACM*, Vol. 28, No. 10, 1985, pp. 1030–1044.
3. D. Chaum and J.-H. Evertse, "A Secure and Privacy-Protecting Protocol for Transmitting Personal Information Between Organizations," *Advances in Cryptology, Proc. Crypto'86, LNCS 263*, A.M. Odlyzko, Ed., Springer-Verlag, 1987, pp. 118–167.
4. D. Chaum, "Blinding for Unanticipated Signatures," *Advances in Cryptology, Proc. Eurocrypt'87, LNCS 304*, D. Chaum and W.L. Price, Eds., Springer-Verlag, 1988, pp. 227–233.
5. D. Chaum, "Privacy Protected Payments: Unconditional Payer and/or Payee Untraceability," *SMART CARD 2000: The Future of IC Cards, Proceedings of the IFIP WG 11.6 International Conference, Laxenburg (Austria), October 19–20, 1987*, North-Holland, Amsterdam 1989, pp. 69–93.
6. D. Chaum, A. Fiat, and M. Naor, "Untraceable Electronic Cash," *Advances in Cryptology, Proc. Crypto'88, LNCS 403*, S. Goldwasser, Ed., Springer-Verlag, 1990, pp. 319–327..
7. D. Chaum, "Online Cash Checks", *Advances in Cryptology, Proc. Eurocrypt'89, LNCS 434*, J.-J. Quisquater and J. Vandewalle, Eds., Springer-Verlag, 1990, pp. 288–293.
8. D. Chaum, B. den Boer, E. van Heyst, S. Mjølsnes, and A. Steenbeek, "Efficient Offline Electronic Checks," *Advances in Cryptology, Proc. Eurocrypt'88, LNCS 330*, C.G. Günther, Ed., Springer-Verlag, 1988, pp. 294–301.
9. D. Chaum, "Achieving Electronic Privacy", *Scientific American*, August 1992, pp. 96–101.

10. D. Chaum and T.P. Pedersen, "Wallet Databases with Observers," *Advances in Cryptology, Proc. Crypto'92, LNCS*, E.F. Brickell, Ed., Springer-Verlag, to appear.

ISO-OSI Security Architecture

Jan Verschuren[1], René Govaerts[2], Joos Vandewalle[2]

[1] TNO, Department EIB,
P.O. Box 5013, NL-2600 GA Delft, The Netherlands
[2] ESAT Laboratory K.U.Leuven,
K. Mercierlaan 94, B-3001 Heverlee, Belgium

Abstract. The Reference Model for Open Systems Interconnection (OSI-RM) enables two APs – residing on different end-systems – to exchange information with each other. In case the information exchanged is transmitted via public telecommunication lines, certain attacks can be envisaged.
Here the OSI-RM is described as well as the attacks threatening the transmitted information.
Subsequently security services are indicated which can protect the communication between two APs. Equipping the OSI-RM with security services makes it possible for APs to exchange information in a secure way. Guidelines are given with respect to the integration of security services in the OSI-RM.

1 Introduction

Formerly, computer systems were mainly stand-alone: a computer was able to process programs (also referred to as application processes (APs)) using its resources. Later computers were coupled to each other by means of a network. The possibility to exchange data between computers was considered to be an advantage. The distributed configuration offered for example the possibility to store certain data at only one location. Besides it became possible to run jobs on other computersystems, if one's computer was not able to do so. To maximize the effect of this configuration it was necessary that all computersystems could be coupled to each other. This goal is not reached if only computersystems of certain manufacturers can communicate with each other. More specifically that would imply that only *closed* communities of computers (that is from the same manufacturer) could communicate with each other in a meaningful way. IBM's Systems Network Architecture (SNA) and DEC's Digital Network Architecture (DNA) are just two examples of communication software packages produced by manufacturers to allow *their* systems to be interconnected together.

Thus standardization was necessary. Efforts of the ISO (International Standards Organisation) resulted in the Open Systems Interconnection Reference Model (OSI-RM) [1]. This model specifies the communication subsystem which enables (application processes on) computersystems of different manufacturers to communicate with each other. This is illustrated by Fig. 1.

Some examples of application processes that may wish to communicate in an open way are:

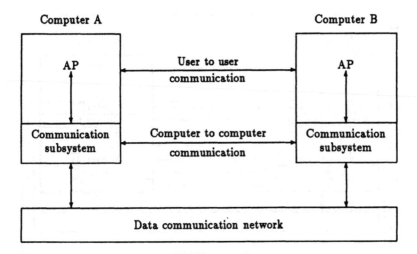

Fig. 1. Computer communication schematic

- a process (program) accessing a remote file system.
- a process controlling a piece of equipment (e.g. a robot) which is linked to a supervisory plant control system. The latter guides the robot control program. In order to do that, communication has to take place between the robot control program and an AP in the plant control system.

2 The ISO Reference Model

The ISO Reference Model for Open Systems Interconnection consists of 7 layers which form together the communication subsystem (Fig. 2).

The layers can be indicated by means of their names or by means of a number N (N = 1,2,...,7). Each layer performs a well defined function in the context of the communication between two APs. It operates according to a defined protocol. (Protocols are sets of rules that govern the exchange of information between two entities.) Performing this protocol results in the exchange of Protocol Data Units (PDUs) between two (protocol) entities that belong to the same layer. PDUs which are exchanged between two protocol entities of layer N are referred to as N-PDUs. These are transported by invoking services provided by the lower layer. More specifically, the N-PDU (which is also named (N-1)-SDU: (N-1) Service Data Unit) is a parameter of a service request which is issued by the N-layer to the (N-1)-layer. Subsequently the (N-1)-protocol entity will feel the need to

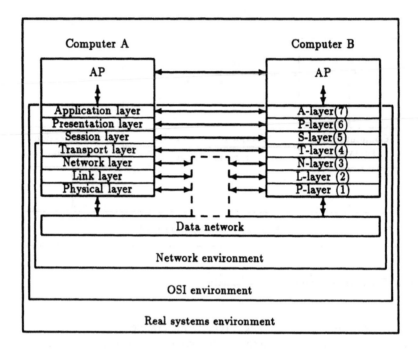

Fig. 2. Overall structure of the ISO Reference Model

exchange an (N-1)-PDU with a peer entity of layer (N-1). The (N-1)-PDU will be equal to the N-PDU concatenated with a so called PCI-block (protocol control information block) which will contain at least the source and the destination address of the service access points of the two communicating protocol entities. This PCI-block governs the PDU-exchange between the protocol entities as it helps the receiving protocol entity to correctly interpret the received PDU. At an (N-1)-service access point ((N-1)-SAP) a protocol entity of layer N can invoke a service provided by a protocol entity of layer (N-1). Figures 3 and 4 illustrate the exchange of PDUs between protocol entities.

The functions of the different layers can be described as follows.

– The physical layer transforms the information to be sent (represented in bits) to (physical) signals which can be transported by the transmission medium.
– The link layer provides the network layer with a reliable information transfer facility. It is thus responsible for such functions as error detection and, in the event of transmission errors, the retransmission of messages.

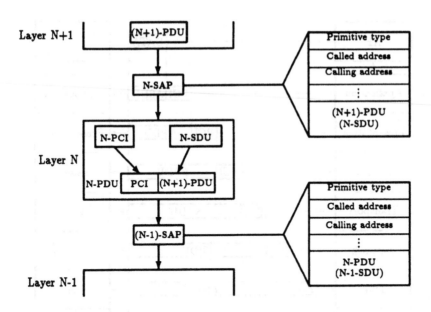

Fig. 3. Interactions between protocol entities in the same system

- The network layer is responsible for the establishment and clearing of a network wide connection between two transport layer protocol entities. It includes such facilities as network routing (addressing).
- The transport layer provides the session layer with a reliable data transfer facility which is *independent* of the type of network which is being used to transfer the data.
- The session layer is responsible for establishing and synchronizing the dialogue between APs. Synchronizing a dialogue means that the dialogue can be resumed from specific synchronization points in case of errors.
- The presentation layer performs the conversion from an abstract syntax (e.g. type character) to a concrete syntax (e.g. ASCII) and vice versa. Figure 5 illustrates this.
- The application layer enables APs to get access to a range of network wide distributed information services. For instance an AP can get access to a remote file server AP which enables the requesting AP to get access to files managed by the file server AP.

As mentioned before, an N-protocol entity has to transmit PDUs (N-SDUs) – which it gets as a parameter in a service request – to a peer protocol entity. Two possible ways exist to do this.

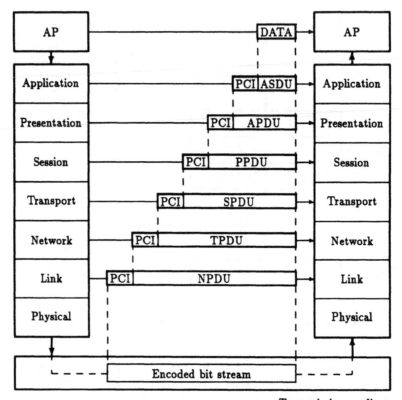

Fig. 4. Interactions between protocol entities in different systems

1. The entity first establishes a connection with the peer entity. If this is settled, then the N-SDU in question is transmitted via the connection. This is referred to as a "connection oriented" service.
2. The N-SDU is transmitted without first establishing a connection. This is denoted by the term "connectionless" service.

Out of the foregoing it turned out that the network layer is responsible for routing PDUs between computer systems. A schematic representation of a computer system is given in Fig. 6.

The nodes must be able to transmit an arriving PDU in the right direction. Therefore a node must be equipped with protocol entities belonging to the lower three layers of the OSI-RM. Figure 7 gives a schematic illustration of a node and the function it performs.

The total address of an AP does not only consist of the physical network-wide address of the computer system it is running on. It is built up out of SAPs in

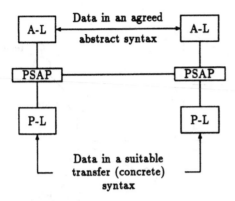

Fig. 5. Function of the presentation layer

the following way.

$$APaddress = PSAP + TSAP + NSAP$$

Here PSAP, TSAP and NSAP denote the addresses of the service access points between the application layer protocol entity to which the AP is connected and the presentation layer, between the session layer and the transport layer and between the transport and the network layer. The NSAP also contains the physical network-wide address of the system in which the AP is resident.

3 Protecting Computer-Computer Communication

3.1 Threats

The fact that information transferred between two computers is transported via a public medium (e.g. the telephone network), makes it extra vulnerable in comparison with information exchanged between APs both running on the same stand-alone system. More specifically data in stand-alone systems is not easily available whereas data transported via public networks is: in general public media are not protected by a physical barrier. Therefore everyone can easily tap information or even modify it. Thus two forms of attacks can be discerned.

- passive attacks resulting in unauthorized disclosure of information.
- active attacks resulting in modification of the transmitted information.

From Sect. 2 it turned out that the information which is transferred via the data communication network is twofold:

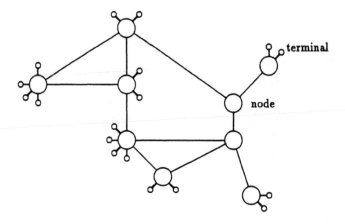

Fig. 6. Computernetwork

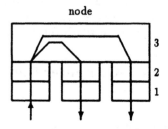

Fig. 7. Routing function of a node

- user data
- data governing the communication between two entities (protocol control information)

So passive attacks can result in the disclosure of user data or of information concerning the communication (e.g. the addresses of the sending and receiving computersystem).

Active attacks intend to modify user data or PCI-blocks. Changing the address of the sending computersystem is an example of modifying the PCI-block. That will make the receiver think that the data came from another system.

...reats just mentioned refer to the situation where two APs are already communicating. Prior to this phase, the communication must be started: an AP attempts to set up a connection with another AP. Then it must be decided if the two APs are allowed to communicate with each other. More specifically it must be prevented that an AP uses resources (e.g. data belonging to another AP) in an unauthorized way.

After the data has been transmitted from sender to receiver, other threats are present:

- the receiver states it did not get the data
- the sender states it did not send the data

So the threats to be envisaged, can be summarised as follows:

- unauthorised access to resources
- disclosure or modification of transmitted data
- false statements of entities which sent or received data

3.2 Protection

Protection needs to be supplied to prevent the threats as mentioned in par. 3.1 to work out successfully. So called security services can realise this protection. The following security services exist.

- access control service; provides protection against unauthorized use of resources accessible via OSI.
- confidentiality service; provides protection against unauthorized disclosure of information.
- integrity service; provides a situation where the receiving entity can deduce whether the information received, is modified.
- authentication service; provides a situation where the receiving entity can deduce if the information received, has been sent by the claimed entity.
- non-repudiation. This security service can take three forms: *Proof of origin*. The recipient of data is provided with proof of the origin of data which will protect against any attempt by the sender to falsely deny having sent the data. *Proof of receipt*. The sender of data is provided with proof of receipt of data which will protect against any attempt by the recipient to falsely deny having received the data or its contents. *Proof of submission*. The sender of data is provided with proof of submission which will prove he has really sent the data.

4 Security Services in the OSI-RM

In the previous section, the threats to user data and to the protocol control information were mentioned. Looking at Fig. 4 it can be derived what information can be protected at what layer. It follows that at the level of the AP only user data can be protected. Security services within the protocol entities of the OSI-RM can protect protocol control information as well. The security services – integrated in the protocol entities – can have different forms. That is mainly dependent on the way the communication is set up between two entities. (More specifically, that can be done by means of a connectionless or a connection-oriented service.)

Now the definition of the security services (with respect to the OSI-environment) will be given (see also [2]).

- Authentication This service can take two forms: peer entity authentication and data origin authentication. The first one refers to a connection-oriented mode of transmitting PDUs, the second one assumes a connectionless mode.
  - Peer entity authentication.
    This service is provided at the establishment of, or at times during the data transfer phase of, a connection to confirm the identities of one or more of the entities connected. This service provides confidence that an entity is not attempting a masquerade or an unauthorized replay of a previous connection.
  - Data origin authentication.
    The service, when provided by the (N)-layer, provides corroboration to the (N+1)-layer that the source of the data is the claimed (N+1)-peer entity. The data origin authentication service is the corroboration of the source of a single connectionless data unit. The service cannot protect against duplication of a data unit.
- Access Control. This service provides protection against unauthorized use of resources accessible via OSI. These may be OSI or non-OSI resources accessed via OSI-protocols. This protection service may be applied to various types of access to a resource (e.g. the use of a communications resource; the reading, writing or the deletion of an information resource; the execution of a processing resource) or to all accesses to a resource.
- Data confidentiality. The following forms of this service are defined.
  - Connection confidentiality. This service provides for the confidentiality of all (N)-user-data on an (N)-connection.
  - Connectionless confidentiality. This service provides for the confidentiality of all (N)-user-data in a single connectionless (N)-SDU.
  - Selective field confidentiality. This service provides for the confidentiality of selected fields within the (N)-user-data on an (N)-connection or in a single connectionless (N)-SDU.
  - Traffic flow confidentiality. This service provides for the protection of the information which might be derived from observation of traffic flows.
- Data integrity. The following forms of this service are defined.
  - Connection integrity with recovery. This service provides for the integrity of all (N)-user-data on an (N)-connection and detects any modification, insertion, deletion or replay of any data within an entire SDU sequence (with recovery attempted; i.e. after detecting that the integrity of the

user data is not fulfilled, subsequent attempts are carried out to get the user data of which the integrity is preserved).

- Connection integrity without recovery. The same as the previous one but with no recovery attempted.
- Selective field connection integrity. This service provides for the integrity of selected fields within the (N)-user data of an (N)-SDU transferred over a connection and takes the form of determination of whether the selected fields have been modified, inserted, deleted or replayed.
- Connectionless integrity. This service provides for the integrity of a single connectionless SDU and may take the form of determination of whether a received SDU has been modified. Additionally, a limited form of detection of insertion or replay may be provided.
- Selective field connectionless integrity. This service provides for the integrity of selected fields within a single connectionless SDU and takes the form of determination of whether the selected fields have been modified.

- Non-repudiation. The three forms this service can take are already defined at the end of par. 3.2.

Part 2 of ISO-standard 7498 [2] gives guidelines at which layer(s) the security services can be provided (Table 1).

Making a choice out of them depends on the security policy of the two APs. A concise explanation will be given on this matter. A security policy specifies how sensitive information has to be protected [3], [4], [5], [6]. Each AP tries to follow a certain security policy when communicating with another AP. If both policies do not agree initially, then negotiation is necessary. If this is successful, then the agreed security policy will result in a set of invoked security services which will be applicable to the communication. More specifically, not all security services available, will be used for each information exchange. Consider for example the following situation: an AP wants to extract address-information out of a file system on a remote end-system. In that case the requesting AP is concerned about the integrity of the received address, not about its confidentiality. So the integrity service needs to be applied whereas the confidentiality service is not necessary.

So dependent on the agreed security policy a set of security services needs to be invoked. For realising the security services so called security mechanisms have to be applied. E.g. the confidentiality service can be realised by the encipherment mechanism. Table 2 gives the relationship between services and mechanisms. The description of all mechanisms is given in [2]. To prevent misunderstanding of the table given in Table 2, the following remark may be helpful. Although encipherment forms only one column of the table, cryptographic techniques may be employed as part of other mechanisms such as digital signature, data-integrity, authentication. An example of a security mechanism which needs not to use cryptographic techniques is routing. The routing mechanism causes information to be transmitted via a special communication path, e.g. via one or more secure subnetworks.

Service	Layer						
	1	2	3	4	5	6	7(*)
Peer Entity Authentication	•	•	Y	Y	•	•	Y
Data Origin Authentication	•	•	Y	Y	•	•	Y
Access Control Service	•	•	Y	Y	•	•	Y
Connection Confidentiality	Y	Y	Y	Y	•	•	Y
Connectionless Confidentiality	•	Y	Y	Y	•	•	Y
Selective Field Confidentiality	•	•	•	•	•	•	Y
Traffic Flow Confidentiality	Y	•	Y	•	•	•	Y
Connection Integrity with Recovery	•	•	•	Y	•	•	Y
Connection Integrity without Recovery	•	•	Y	Y	•	•	Y
Selective Field Connection Integrity	•	•	•	•	•	•	Y
Connectionless Integrity	•	•	Y	Y	•	•	Y
Selective Field Connectionless Integrity	•	•	•	•	•	•	Y
Non-repudiation, Origin	•	•	•	•	•	•	Y
Non-repudiation, Delivery	•	•	•	•	•	•	Y

Key: Y: Yes, service should be incorporated in
the standards for the layer as a provider option.

•   Not provided.

*   It should be noted, with respect to layer 7, that the
application process may, itself, provide security services.

Table 1. Illustration of the relationship of security services and layers

To give an idea of the effect of placing a security service at different layers, examples with respect to three security services will be given.

Confidentiality. Encipherment at two different layers is considered.

- application-layer
- physical-layer

As was described in Sect. 2, the nodes of the network are equipped with protocol entities belonging to the three lower layers of the OSI-RM. This implies that encipherment of the information at the application-layer does not cause problems: the nodes can read and interpret the address information in the PCI-

Service \ Mechanism	Encipherment	Digital Signature	Access Control	Data Integrity	Authentication Exchange	Traffic Padding	Routing Control	Notarization
Peer Entity Authentication	Y	Y	•	•	Y	•	•	•
Data Origin Authentication	Y	Y	•	•	•	•	•	•
Access Control Service	•	•	Y	•	•	•	•	•
Connection Confidentiality	Y	•	•	•	•	•	Y	•
Connectionless Confidentiality	Y	•	•	•	•	•	Y	•
Selective Field Confidentiality	Y	•	•	•	•	•	•	•
Traffic Flow Confidentiality	Y	•	•	•	•	Y	Y	•
Connection Integrity with Recovery	Y	•	•	Y	•	•	•	•
Connection Integrity without Recovery	Y	•	•	Y	•	•	•	•
Selective Field Connection Integrity	Y	•	•	Y	•	•	•	•
Connectionless Integrity	Y	Y	•	Y	•	•	•	•
Selective Field Connectionless Integrity	Y	Y	•	Y	•	•	•	•
Non-repudiation, Origin	•	Y	•	Y	•	•	•	Y
Non-repudiation, Delivery	•	Y	•	Y	•	•	•	Y

Key: Y: Yes, the mechanism is considered to be appropriate,
either on its own or in combination with other mechanisms.

• The mechanism is considered not to be appropriate.

Table 2. Relationship of security services and mechanisms

block added by the N-layer as this is not enciphered. (The same counts for encipherment at the level of the T-layer.) Encipherment applied at the transport or application layer is referred to as end-to-end-encryption as during transport no deciphering of the PDUs takes place; they are deciphered for the first time when they arrive at their destination. Encipherment at the physical-layer necessitates decipherment of the PDU at the nodes to enable the node to determine the direction in which it has to send the PDU. This implies that at the node the whole PDU (including the user data sent by the AP) is in the clear. This may be unacceptable for the communicating APs. Encipherment at the physical layer has an advantage however. It will imply that an enemy who taps the line will see nothing of the structure of the information. He will therefore be unaware of the sources and destinations of messages and may even be unaware whether messages are passing at all. This provides 'traffic-flow security' which means that not only the information, but also the knowledge of where the information is flowing and how much is flowing is concealed from the enemy. By implementing encipherment at the lowest level, traffic-flow security can be obtained.

Access Control and Authentication. In Sect. 2 the address structure was seen. APaddress = PSAP + TSAP + NSAP. As a result of this, it can be said that the access control service at the A-layer can deny or approve access to a specific AP as at that level the AP is fully specified. At the N-layer (or T-layer) however, only access can be given or denied to a group of APs. Analogously, the data-origin and peer-entity authentication services realised in the N-layer or T-layer can only assure that the data origin respectively peer-entity is a member of a group of entities. Realization of the mentioned security services in the A-layer gives assurance with respect to a *specific* entity.

5 Conclusions

The OSI-RM enables two APs – residing on different end-systems – to exchange information with each other. Equipping the OSI-RM with security services, makes it possible for APs to exchange their information in a secure way. With respect to the realisation of security services into the OSI-RM, a lot of possibilities exist. Part 2 of ISO standard 7498 [2] gives guidelines for integrating security services in the OSI-RM. However concerning further, more detailed implementations of the security services in the OSI-RM, few standards are available.

References

1. Open Systems Interconnection Reference Model, Part 1: Basic Reference Model, ISO 7498-1 (CCITT X.200). Melbourne 1988.
2. Open Systems Interconnection Reference Model, Part 2: Security Architecture, ISO DIS 7498-2, July 19, 1988.
3. Landwehr, Carl E., "Formal Models for Computer Security," *Computing Surveys*, Vol. 13, No. 3, September 1981, pp. 247-278.

4. Bell, D. Elliot and LaPadula, Leonard J., Secure Computer Systems: Unified Exposition and Multics Interpretation, MTR 2997 rev.1, The MITRE Corporation, March 1976.

5. Biba, K.J., Integrity Considerations for Secure Computer Systems, MTR-3153, The MITRE Corporation, June 1975; ESD-TR-76-372, April 1977.

6. Clark, D.D., Wilson, D.R., "A comparison of Commercial and Military Computer Security Policies," *Proceedings of the 1987 IEEE Symposium on Security and Privacy*, IEEE Computer Society Press, April 1987.

Security Aspects of Mobile Communications

Klaus Vedder

GAO Gesellschaft für Automation und Organisation mbH*
Euckenstr. 12, 8000 München 70, Germany

Abstract. Security requirements and services of a mobile communication system differ, due to the radio communication between the user
and the base station, extensively from those of a fixed network. There
is no physical link in the form of a (fixed) telephone line between the
user and the local exchange, which could serve to "identify" the user for
routing and charging purposes. His identity has to be verified over an air
interface. Authentication by means of cryptographic procedures is thus
required to stop impostors from taking on the identity of somebody else
and "transferring" calls and charges. Eavesdropping on the radio path
and intercepting a conversation or data or tracing the whereabouts of a
user by listening to signalling data are other serious threats. This paper
discusses the countermeasures designed into one of the most advanced
radio networks, the Global System for Mobile Communications, as well
as some security aspects of the network management related to these
measures. Some of the differences between this system and the planned
Digital European Cordless Telephone system are highlighted.

1 Introduction

The Global System for Mobile Communications (GSM) is a digital, cellular radio
system involving, at present, 17 European countries with 24 network operators
and Australia as the first non–European country. By the summer of 1992 several
network operators will have started commercial operation. The immense task of
specifying such a system across all national, company and individual interests
started 10 years earlier in 1982 with the formation of the Groupe Spécial Mobile
by CEPT, the European Conference of Postal and Telecommunications Admin-
istrations. In 1989 the Groupe Spécial Mobile became a Technical Committee of
the newly founded European Telecommunications Standards Institute (ETSI).
In the ten years of its existence, nearly 200 specifications where issued ranging
from functional descriptions covering security aspects to type approval tests of
mobile stations.

One of the fundamental features of this pan–national system is the possibility
to roam across national borders. A user is not restricted to his home network
or his home country. A mobile station can be taken to and used in any of the

---

* A member of the Giesecke & Devrient group of companies

participating countries for making calls subject only to a roaming agreement between the operators of the home network and the visited network. The implications of this for the network management, the billing procedures and the security services are manifold. Authentication and billing are done by the operator of the home network, the so-called Home Public Land Mobile Network (HPLMN). There is, therefore, no need for a standard authentication algorithm common to all networks; every operator can run his own proprietary algorithm. Only the interface parameters are specified systemwide. The data needed for checking the authenticity of a user is usually generated by the home network. The actual verification and thus the access control to the system are handled locally by the Visitor Location Register (VLR), where the subscriber is temporarily registered. This may be a VLR of a visited network. For details of the GSM network architecture the reader is referred to [8] which also contains a list of definitions of the system elements.

When travelling the subscriber need not even take his mobile station (MS) with him. It suffices to take his subscriber card, the so-called Subscriber Identity Module (SIM), and to insert this into any mobile equipment (ME). The ME is the MS without the SIM. The SIM contains all the necessary information about the subscription such as the International Mobile Subscriber Identity (IMSI) as well as the network specific authentication algorithm and the secret, subscriber specific authentication key. No subscription related information is contained in an ME. This split of a mobile station into a radio part and a subscription part gives the network operator, on whose behalf the SIM has been issued, the complete control over all subscription and security related data. The SIM is thus an integral part of the overall security system of each and, therefore, all networks.

Mobile communication has to cope with security issues unknown to a fixed network. Listening to the radio link between the mobile station and the corresponding base station is easily done and can hardly be prevented. How useful, if at all, the intercepted data are to the eavesdropper depends on the security services provided by the network operator. One of the novel features of GSM is the enciphering of this link to protect user and signalling data. A special cipher has been developed for this purpose. It is integrated into the mobile equipment as a dedicated piece of silicon. The cipher key is derived by the SIM as part of the authentication process. So the enciphering can only be activated after the identity of the user has been (successfully) verified. To run the authentication procedure the mobile station is, of course, required to send the identity of the user over the air interface. To counteract the threat of tracing the whereabouts of a user, temporary identities are issued by the VLR.

After the discussion of the security features and the security services provided in a GSM network in the following sections, we take a close look at the role played by the Subscriber Identity Module as a secure device for storing keys and algorithms. This is followed by key management issues and a list of definitions and abbreviations. This paper is mainly concerned with the security aspects and features of the introductory phase of GSM (Phase 1). Phase 2, which is supposed to begin in 1993 and will offer the user additional functionality, will also see some changes with respect to the security services. Some of the anticipated changes are highlighted. For a general description and background information on the GSM system the reader is referred to [4] and [12].

The security services specified for the new Digital European Cordless Telephone system (DECT) are similar in nature to those provided by GSM. One of the main differences is that roaming is not an integral part of this cordless system and that, therefore, the authentication process may have to be done by a visited network. For this reason a DECT Standard Authentication Algorithm (DSAA) has been specified by ETSI. Furthermore, the DECT Authentication Module, which plays a role similar to the SIM, is only optional. A very detailed threat analysis of the DECT system as well as the specification of the authentication functions is contained in [11] which forms one of the 11 parts of the DECT specification. For a discussion of the security features of DECT the reader is referred to [20]. The threat analysis given in [11] applies to a large extent to any mobile communication system.

2   Security Threats and Features

The radio path and the access to the mobile services are two areas where a mobile communication system does, naturally, not provide the same level of protection as a fixed network unless additional security measures are taken. Two basic threats are the interception of data on the air interface and the illegitimate access to a mobile service.

The interception of user data or signalling information related to the user may result in a loss of confidentiality of this data or the loss of the confidentiality of the user's identity with respect to the outside world. (There is no anonymity within the system as for instance offered by a pre–paid telephone card. However, it would be comparatively easy to manipulate a mobile so that the charges are not deducted from the card. This would not be noticed by the network, as the signalling channel cannot be used to send an acknowledgement from the card to the network.) User data is transferred over traffic channels as well as over signalling channels. The signalling channel carries, apart from obvious user related signalling information elements such as the called and the calling telephone numbers, user data in form of short messages. This service allows the user to receive and send messages of up to 160 bytes over the radio link without activating a traffic channel.

The illegitimate use of a service is not only of concern with respect to proper billing. It is clearly important that billing is always possible and that only the subscriber, who has caused the charge, is billed for it. The not so obvious illegitimate use is masquerading. Impersonating a subscriber and claiming afterwards that this subscriber (or to be more precise his subscriber card) must have been in a particular place at a particular time is certainly not a very widespread threat but one which could prove very serious indeed in certain circumstances. Due to the cell structure of the network and the constant updating of the location information of a SIM, which is necessary for a proper and timely delivery of mobile terminated calls, the network knows the location of a subscriber card down to the cell (of course only if the SIM is inserted into the mobile and this is switched on). The area covered by a cell depends on the location of the cell and ranges from a few hundred metres to about a 35 km radius around the base station.

To protect network operators and subscribers against such attacks inherent in any unprotected radio link and an uncontrolled access to a mobile service, the *implementation* of the following security features is mandatory in any GSM network on both the fixed infrastructure side and the MS side.

- subscriber identity confidentiality
- subscriber identity authentication
- user data confidentiality
- signalling information confidentiality

The functional description of these security features is given in the Technical Specification GSM 02.09 [6]. The protection of the mobile station itself, that is to say the SIM, against unauthorised usage is part of the functionality of the SIM [7]. This will be discussed in Sect. 4.

One security service not provided for by GSM is the authentication of the network by the mobile station. This is quite an important feature of the DECT system. It is required due to the different environment a DECT portable part is used in and the possibility to change subscription data or even to block a subscription over the air interface (see [11], [20] for details).

3 The Security Services

A functional description of security features is by its very nature, due to differing interpretations and implementations, not sufficient to ensure that interoperability between networks and the same level of security are achieved throughout GSM. The specification of the network functions needed to provide the security services for the features listed above are contained in the Technical Specification GSM 03.20 [9]. This document also contains the external specification for the cryptographic algorithms required to perform these services. The algorithms themselves are, except for the cipher algorithm(s), network operator matter.

From a user point of view it is not relevant whether the user–related data to be protected is contained in a traffic or a signalling channel. We may therefore say that GSM basically provides three security services:

- temporary identities for the confidentiality of the user identity,
- authentication for the corroboration of the identity of the user,
- enciphering for the confidentiality of user-related data.

In the following three sections we will discuss each one of them in turn.

3.1 Temporary Identities

Before a user can make, say a call or go on standby for receiving calls, his identity has to be known to the network. Rather than sending the International Mobile Subscriber Identity (IMSI), which uniquely identifies the subscriber worldwide, a temporary identification number is sent in most instances.

The purpose of temporary identities is to deny an intruder the possibility of gaining information on the resources a subscriber is using and to prevent the tracing of the user's location. In addition, matching the user and the data transmitted is made more difficult.

To achieve this "the IMSI is not normally used as an addressing means on the radio path" and "should be used only when necessary" [9]. Clearly, the IMSI has to be used for the set up of a session if there are no other means to identify a mobile subscriber. This is for instance the case when the subscriber uses his SIM for the first time or at a data loss in the VLR where the subscriber is temporarily registered.

When the SIM is used for the first time, the MS will read the default Temporary Mobile Subscriber Identity (TMSI) stored in the SIM at pre–personalisation (see Sect. 5.3) and send this value to the VLR. As this TMSI is unknown to the VLR, the VLR will request the IMSI from the MS. It then assigns a TMSI to the subscriber and transmits this identifier (after a successful authentication) and the activation of the cipher process in an enciphered form to the MS. The MS deciphers the data and stores the TMSI and information about the present location in the SIM. From then on this TMSI will be used by the MS instead of the IMSI until a new TMSI has been assigned to the subscriber.

Though the TMSI consists of only 5 digits, the subscriber is uniquely identifiable. For the TMSI is unique within the location area where the MS moves, and the location area identification (LAI) is always used in conjunction with the TMSI. To be able to identify and locate the subscriber, the TMSI is stored together with the IMSI and the LAI in the VLR.

A new TMSI is to be assigned at each location update procedure. GSM 03.20 specifies six scenarios for the allocation of a new TMSI. If there is no malfunctioning in the system, the IMSI will never again be used for call set up. Even if the SIM has moved to a new VLR in a different network, the new VLR can and must obtain the IMSI from the old VLR by using the old TMSI and LAI which have been sent by the mobile station.

3.2 Authentication

Authentication is the corroboration that an entity is the one claimed [14, part 1] or the verification of the identity of the SIM.

The purpose of subscriber identity authentication is "to protect the network against unauthorized use" [9] and thus to ensure correct billing and to prevent masquerading attacks.

The General Procedure. Authentication is in the domain of the network operator and every operator may use his own algorithm(s). A proposal for a possible algorithm is available upon appropriate request [9]. However, to achieve interoperability the authentication protocol and the lengths of the parameters (input, output and key) as well as the algorithm execution time have been specified.

On the network side the authentication algorithm, which is denoted by A3, is implemented in the HLR or the Authentication Centre (AuC), while on the MS side it is contained in the SIM. As an option A3 may be implemented in VLRs (see Sect. 3.2).

The method employed between the HLR/AuC and the SIM is a challenge-response mechanism using "non–predictable numbers". Figure 1 shows the authentication of the SIM by the network side. To run this procedure the network has to know the (temporary) identity of the subscriber since his (specific) authentication key K_i is used.

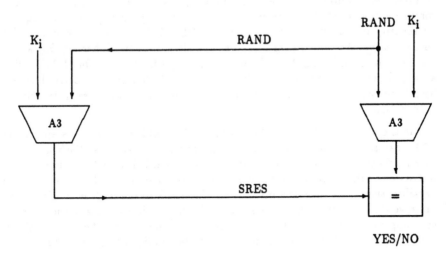

Fig. 1. Authentication procedure

The network side transmits a non–predictable number RAND to the MS as a challenge. The SIM computes the response SRES to RAND by using the algorithm A3 with RAND and the secret subscriber specific authentication key K_i which is stored in the SIM, as input data. SRES is transmitted to the network side. There it is compared with the value computed by the network, which has used the same algorithm with the same RAND and the key associated with the identity claimed by the subscriber. Only if the two values are equal is the MS granted access to the network. For it can be assumed that the SIM is in possession of the right subscriber key K_i and that, therefore, its identity is the one claimed.

Handling of the Parameters. The handling of the parameters is of importance with respect to both security and the efficient dealing with a request by a subscriber.

The parameters have the following lengths:

$$K_i: 128 \text{ bits}; \text{ RAND}: 128 \text{ bits}; \text{ SRES}: 32 \text{ bits}, K_C: 64 \text{ bits},$$

Where K_C is the cipher key used for the ciphering of the air interface. The run–time of the algorithm A3 shall be less than 500 ms.

The verification of SRES is done in the VLR where the MS is currently registered, while the computation is carried out in the HLR/AuC of the home network of the subscriber. The VLR obtains pairs consisting of RAND and the corresponding SRES from the HLR/AuC (of the home network) upon request for security related information using the IMSI of the subscriber who is to be authenticated. Accompanying these pairs is always a new cipher key K_C which has been computed using K_i and the same non–predictable number RAND with an algorithm called A8 (see 3.3). The challenge RAND and the derived values SRES and K_C are called an authentication triplet or a set of security related information.

As the VLR and the HLR/AuC may be thousands of kilometers apart, the VLR may store, as does the HLR, five authentication triplets. Each authentication triplet is used only once and must be discarded after being used. Both restrictions will not apply to Phase 2. The re–use of security related information in failure situations such as a breakdown of the link to the HLR is considered to improve the security level. In Phase 1 the VLR may in such a situation permit outgoing calls without further authentication if the MS has been successfully registered and authentication triplets cannot be obtained from the HLR.

When the user has moved to a different VLR, this new VLR will normally request the IMSI from the previous VLR by sending the old TMSI and LAI (see Sect. 3.1). In either phase the old VLR transfers, together with the IMSI, any (unused) triplets to the new VLR. This speeds up the authentication procedure as the new VLR can, of course, only send a request for triplets to the HLR/AuC after it has learned of the real identity of the subscriber which is through this request to the old VLR.

Option. In Phase 1 the HLR/AuC may transmit the secret subscriber key upon request for security related information to the VLR for the local generation of the triplets used for authentication and enciphering. It is however recommended to restrict this procedure to the HomePLMN [9, clause 3.3.2].

Using this possibility would certainly improve the availability of the service in peak times and, in particular, in situations where the link between the HLR and the VLR is not available. The security implications are, however, severe. The (secret) algorithms A3 and A8, which are used to generate SRES and K_C have to be implemented in the VLR in a secured environment. The other problem is sending secret keys over insecure channels. The links between the VLRs and the HLR are natural places for an attacker to collect IMSIs and corresponding keys. The keys should not be transmitted in clear over the fixed network or the microwave connecting the HLR with the VLRs. In quite a few countries, however, sending enciphered data over such lines is not allowed.

If this method was used for roaming the network operator would have to disclose both algorithms to another operator (or to supply him with black boxes).

Because of the sensitive nature of the key K_i the Technical Committee (the new name of which is SMG for Special Mobile Group as it is now responsible not only for GSM) has approved the deletion of this option from Phase 2.

The use of public key cryptography (see [18] for a summary) would allow the local verification of the response without any *secret* information having to be divulged to the VLR. Authentication mechanisms using public key techniques are also about to be standardised [14]. Some of the reasons, why the classical solution had been opted for both DECT and GSM, are the time constraints of the authentication process and the amount of data to be handled. Changing now to a public key algorithm for the authentication process would be very difficult indeed as the air interface does not support the transmission of the required amount of data.

3.3 Enciphering

The purpose of this security service is to ensure the privacy of the user information on both traffic and signalling channels and of user–related signalling elements on the radio path. End–to–end enciphering is outside the scope of GSM.

The activation of this service is controlled by the network side. It is started by the base station by sending a "start cipher" command to the MS.

A standard cipher, which is denoted by A5, is used throughout the system. This (secret) algorithm, which can be implemented using about 3,000 transistors [3], is contained as a dedicated piece of silicon in the Mobile Equipment and the Base Station, which is the counterpart to the MS on the radio path. For Phase 2 it is intended to have eight instead of one cipher algorithm. However, not all of these have, however, to be supported by every MS.

For several reasons it is not possible that a network operator uses its own proprietary algorithm. For instance, roaming allows calls which originate and/or terminate in a "foreign network". As the ciphering would have to be between the SIM and the home network, routing say a "local" call in a foreign network via the home network would not only be prohibitively expensive but would also create both immense technical and political problems.

A form of a stream cipher is used to encipher the layer 1 data. The plain text is organised into blocks of 114 bits as this is the amount of data which is transmitted during a time slot. The key stream, which is the sequence of bits to be XORed (modulo 2 addition, denoted by $\oplus$) with the data block, is produced by the algorithm A5 as an output block of 114 bits. For synchronisation and other implementation details the reader is referred to [9]. Security aspects of speech communication in general are extensively discussed in [2].

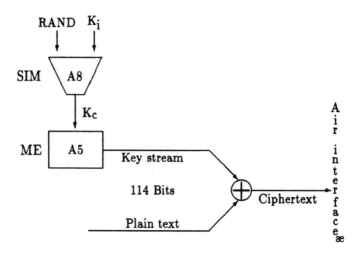

Fig. 2. Cipher key generation and enciphering

The key K_c, which controls the generation of the key stream by the algorithm A5, is derived in the SIM during the authentication process using the network operator specific algorithm A8. A proposal for an algorithm A8 is available on appropriate request [9]. The input for the cipher key generator A8 are the challenge RAND (128 bits) and the subscriber authentication key K_i (128 bits) used also for the authentication procedure, while the output is the key K_c consisting of 64 bits.

Binding the generation of K_c to the authentication process has the following advantage apart from not requiring additional input data. For instance, to illegitimately use a service it is not sufficient just to "bypass" the authentication procedure by, say, manipulating the comparison of SRES in the VLR. Since in this case the MS and the base station would use different cipher keys resulting in an undecipherable garbled message.

It is worth noting that, unlike the symmetric and public key algorithms used for the authentication mechanisms standardised in [14], the authentication algorithm A3 as well as the cipher key generator A8 compress the non–constant input data RAND from 128 bits to 32 and 64 bits, respectively. Even if A3 and A8 are one and the same algorithm, which is feasible judging by the input parameters, a compression takes place. This implies that more than 1 input value can produce the same output and the algorithm is clearly irreversible. The input is, therefore, not derivable from the output even if the key K_i is known. This means that the SIM cannot be used for enciphering or deciphering data.

4 The Subscriber Identity Module

The Subscriber Identity Module (SIM) is a security device which contains all the necessary information and algorithms to authenticate the subscriber to the network. It adds a new dimension of mobility to the subscriber as it is a removable module and may be used with every mobile equipment. The functionality of the SIM is described in GSM 02.17 [7] while its interface to the ME is specified in GSM 11.11 [11] (every SIM has to work with every mobile!).

To achieve its main task of authenticating the subscriber to the network side, the SIM contains a microcomputer with on–board non–volatile memory. In other words, the SIM is a smart card. This card comes in two formats. The IC card SIM is the size of a credit card, while the Plug–in SIM is "obtained" from this by cutting away excessive plastic and thus reducing the size to 25mm by 15mm (see Fig. 3). For the exact dimensions and the location of the contacts the reader is referred to [10] and [13, part 2]. The latter is intended to be used mainly with handheld mobiles which are too small to support an IC card SIM. A third format, having the dimensions 66mm by 33mm, has been specified for the DECT Authentication Module. The electrical and mechanical interfaces are, with the obvious exception of the two small cards, in line with the relevant international standards for IC cards [13]. In some instances, however, more stringent conditions were agreed upon to cater for the needs of the environment the SIMs are used in. These include, for instance, the temperature range the card has to satisfy and the power consumption of the microcomputer. General aspects of the SIM are discussed in [15] and [17].

The microcomputer consists of a CPU and three types of memory. The masked programmed ROM usually contains the operating system of the card and the security algorithms A3 and A8. The RAM is used for the execution of the algorithms and as a buffer for the transmission of data. The non–volatile memory, which is needed to store subscriber specific data such as K_i and IMSI, is EEPROM (Electrically Erasable Read Only Memory). EPROM, which can

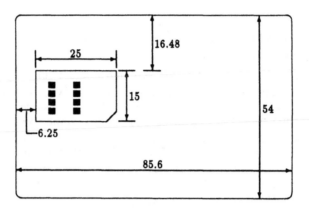

	typical	maximum
ROM	4–6 kByte	16 kByte
RAM	126–160 Byte	256 Byte
EEPROM	2–3 kByte	8 kByte

Fig. 3. SIM formats and memory provided

be written to only once, cannot be used as network and subscriber related information such as TMSI, LAI, abbreviated dialling numbers and short messages are updated frequently or have to be changeable by the subscriber. The memory space offered by present day smart card chips is given in Fig. 3. For more information on smart cards the reader is referred to [19]. Basic security aspects of microcomputers and their manufacture can be found in [16].

The operating system of the card controls the access of the outside world, which may be a mobile equipment or any other interface device, to all data stored in the card. Access is mainly by reading or updating the respective memory cells. Each instruction is, from a security point of view, equivalent to either READ or UPDATE. GSM has specified five (independent) access conditions. These are NEVER, ALWAYS, ADM/AUT, PIN and a second PIN for Phase 2. The access condition ALWAYS means that no security restriction holds. The IC Card Identification number, which identifies the SIM and is also printed on the card itself, may always be read over the interface but will NEVER be permitted to be updated. The definition and use of ADM/AUT is up to the discretion of the operator. This may be one of the other four access condition or a specific procedure which can only be executed by an appropriate administrative authority. PIN and PIN2 are discussed in the following section.

The interface to the outside world consists of eight contacts two of which are reserved for future use. One of the remaining six contacts is needed for cards requiring an external Programming Voltage. All SIMs derive this voltage from the Supply Voltage (V_{CC}) by means of an internal charge pump. The remaining contacts are used for the Reset of the chip, the Clock to drive the chip, the Ground and only one contact for Input and Output. The latter certainly restricts the data throughput but is of an advantage from a security point of view. The typical Baudrate is 9,600 bits/sec which gives a (theoretical) upper bound for the card throughput of, say, a cipher of 3200 bits/sec. This is due to the half duplex transmission protocol and the overhead of having to send 12 bits for every byte. Such a rate is certainly not sufficient for ciphering the data going over the air interface.

4.1   Access to the SIM

Similar to some banking cards, user access to the SIM is controlled by a Personal Identification Number (PIN). A major difference is, however, that the PIN can be freely chosen by the user within the range of 4 to 8 digits and that the user may change the PIN as often as he/she feels necessary. This is possible since the reference PIN is stored in the EEPROM of the chip and the comparison of the value presented by the user with the reference PIN is done in the CPU of the SIM; the PIN does not leave the chip. To protect the user against trial and error attacks, the microcomputer controls the number of *consecutive* false PIN entries. After three such entries the card will be blocked and refuse to work, even if it has been removed in between attempts or a different mobile equipment is used.

A blocked SIM does not even send its identity in form of the TMSI or IMSI to the ME as these data fields are protected against reading by the security level PIN. Checking the PIN prior to the unilateral authentication of the MS by the network side saves signalling time and resources in the network(s) if a false PIN is entered. It does not impair the security of the subscriber as the SIM has no means of authenticating the ME or the network and PINs collected by a fake ME can only be used in conjunction with the genuine SIM.

The user may "unblock" a SIM by presenting the so–called Personal Unblocking Key (PUK) to the card together with a new PIN. The PUK is simply another identification number consisting however of eight digits. As the PUK cannot be changed by the user it could also be stored in the home network and given, if the necessity arises, to the user under mutually agreed security conditions such as the delivery of a special password by the user. It should be recalled that the access of a SIM to a mobile service can easily be suspended by blacklisting the subscription in the HLR/AuC. After 10 consecutive wrong PUK entries the SIM is permanently blocked.

Another novel feature is the disabling of the PIN check. The network operator or the service provider, if the network operator so decides, may set a flag in the card which allows the user to switch off (and on) the PIN check. If the PIN check has been disabled by the user, the access control to the SIM and thus the network (if the card is not blacklisted) is based merely on the possession of the card. It remains to be seen whether subscribers will be allowed to use this feature.

With the introduction of new services in Phase 2 the PIN will be the means to control user access to the actual GSM operation instead of the GSM application, while certain optional data fields may be protected by a second PIN and PUK. For instance, *Advice of Charge* is intended not only as an advice to the customer about the number of units spent but may also be used for billing purposes. Due to worldwide roaming the issuer of such a special card has not necessarily the latest billing information. What has been said about prepaid cards also holds for such cards. As these cards will, however, mainly be used in closed user groups and not on an anonymous basis, manipulating a mobile to suppress charging information has implications for the user. Resetting the charge counter should clearly not be under the control of the user and thus not under the control of the PIN. For this reason PIN2 and PUK2, which control the access to this and similar data fields independently of PIN and PUK, have been introduced for Phase 2. The same boundary conditions as for PIN and PUK apply to PIN2 and PUK2.

5 Key Management

In this section we consider some of the aspects concerning the authentication centre (AuC) and the handling of subscriber authentication keys. The keys K_i have to be very well protected to avoid charging the wrong subscriber and to counteract other threats posed by impostors.

5.1 The Authentication Centre

GSM 03.20 states that the subscriber authentication key K_i is allocated, together with the IMSI, at subscription time and that K_i is stored in an Authentication Centre (AuC). The functionality of the authentication centre is not described in any of the specifications. It is of concern only to network operator and is thus dependent on the specific security and administrative requirements of the operator. It is only stated that the authentication algorithm(s) A3 and the cipher key generating algorithm(s) A8 are implemented in the AuC.

The AuC is the heart of the security. A malfunction or a temporary loss of the information contained in it would have severe consequences for the security as it affects the generation of the authentication triplets. Since other information about the subscriptions including possibly black lists of barred subscriptions is contained in the HLR, it is only logical to "integrate" the AuC into the HLR. In networks with more than one HLR the back-up and overload facilities could be distributed over several HLR/AuCs.

Key management is a major issue when designing an AuC. The number of subscriptions is estimated to reach 20 million in the year 2000 in Europe. The larger operators will contribute several millions each to this number. The method used for generating and storing potentially several million subscriber authentication keys and the handling of the authentication requests is of importance for both the secure and the smooth running of the network.

5.2 Key Generation

There are two standard methods for generating keys. This may be done by using a random number generator or by means of an algorithm which is used to derive the key from user related data under the control of a master key. Which of the two methods or variation thereof the designer of the security system chooses depends on the local circumstances. Both methods have their advantages and disadvantages which we will briefly discuss in the light of the boundary conditions given in the Technical Specification GSM 03.20 [9].

Deriving a key. The main advantage of deriving a key from non–secret (subscription) data under a master key MK is that such derivable keys need not to be stored and that the back–up of the subscriber keys is reduced to the back–up of the master key. No data banks containing secret information are thus required in the AuC nor at the back–up facility. When an authentication request comes from the VLR via the HLR, the AuC would just load the relevant data into the algorithm and derive the subscriber authentication key K_i from this data using the top secret master key MK. K_i would then be loaded with the random number into A3 and A8 for the generation of SRES and K_C.

The main problem is of course to keep the very secret key MK secret. Anybody coming into possession of this key could, if he also knows the secret algorithms A3 and A8, compromise every card issued under MK. One could reduce the potential damage by replacing the master key periodically. As a consequence, more secret keys have to be maintained and a logical link has to be established between the respective master key and the subscription (this could be done by coding the key number as part of the IMSI). The number of master keys being used at the same time depends on the length of time each one of them has been employed for generating subscriber authentication keys as well as on the validity period of those SIMs.

The selection of the algorithm, which is used for deriving K_i, and the input data to go into this algorithm depend on several boundary conditions. These include the length of the key K_i (128 bits) and whether one of the algorithm(s) already available in the AuC is suitable or whether a specific algorithm should be employed. Considering the number of authentication requests the algorithm has to be fast if one wants to avoid "queues" or having too many secured boxes running in parallel.

A natural candidate for the input data would be the IMSI which consists, however, of only 15 digits each one coded on half a byte [10]. A natural candidate for the algorithm is the DEA [1] which is also available in hardware. As the key K_i consists of 128 bits and the DEA has an input block of 64 bits one has to "expand" the IMSI to 16 digits and apply the DEA twice. To improve the distribution of the derived keys one could first reduce the length of the IMSI to 10 digits by removing the 5 digits which are identical for all IMSIs in the HPLM and use parts of other subscription related data for the remaining 6 digits. Two possibilities to obtain K_i from this value UD representing the user data are as follows, where $\|$ denotes the concatenation of the two expressions:

$$K_i = DEA_{MK_{LEFT}}(UD) \| DEA_{MK_{RIGHT}}(UD)$$

or

$$K_i = DEA_{MK}(UD) \| DEA_{MK}(DEA_{MK}(UD)).$$

In the first case the master key consists of two parts of 64 bits each, while it has only 64 bits in the second case.

A *variation* of this method would be to use as the input data UD the IMSI or parts thereof together with a string of random bits. This is however somewhat counterproductive to the main reason for choosing the method in the first place. For this random data has to be stored, though not enciphered, against the IMSI in a data bank in the AuC. On the other hand, this variation has the advantage that the random data is not publicly available and not related to the subscription. A compromise of the master key does, therefore, not automatically impair the security of the whole system. One could even go a step further and use only random data for the input to the algorithm.

The key as a random number. Using a random number generator to produce the subscriber authentication keys insures that all strings consisting of 128 bits are equally likely. This cannot be achieved by an algorithm using IMSIs as an input.

The main difference is, however, that there is no natural link between the subscription and the authentication key. This requires all keys to be stored against some subscription specific data in a data bank of the AuC and to be backed-up at a physically different location. As the authentication request involves the IMSI this would again be a natural choice. To protect the keys against unauthorised reading in the AuC they have to be stored in an enciphered form. The key or keys used for deciphering the subscriber authentication keys is clearly very sensitive. A compromise of such a key is in itself not as much a security breach as a compromise of the master key used to derive the authentication keys from subscription data. The attacker also needs a dump of the data bank.

Similar things as before can be said about the choice of this algorithm. Assuming this to be the DEA in electronic code book mode, we can also see that the times required for providing a key for the authentication request are about the same (assuming that access to the data bank causes no significant overhead). In both instances two DEA encipherments or decipherments have to be executed.

5.3 Pre–personalisation

"Pre–personalisation is assigning and loading a SIM with authentication key and IMSI, and is done using a pre–personalisation key" [7]. In general, no subscriber related information is required in the SIM for the access of a GSM service. A pre–personalised SIM may thus contain all information necessary for the GSM operational phase and be ready for use subject only to its 'release' in the HLR/AuC.

This certainly facilitates the handling of SIMs and the corresponding PIN- and PUK-mailers.

Which of the methods described in Sect. 5.2 is employed for the generation of the subscriber keys depends also on the administrative environment of the pre–personalisation. Deriving K_i from subscription data, which is known prior to the pre–personalisation of the SIM, allows the computation K_i independently in the HLR/AuC and at pre–personalisation time. K_i need, therefore, not be transmitted between the two places. The administrative problems of this method should, however, be considered with care. If K_i is a random number or depends partially on random data, then it may be generated in the HLR/AuC or at the place of pre–personalisation. K_i and, if the latter is the case, the random data have to be transmitted in a secure way between the two entities. Both solutions have their advantages and disadvantages and a decision for one or the other should not be taken without considering all security and administrative boundary conditions.

Abbreviations

This section is meant to give the reader a quick look–up table containing the abbreviations used in this paper together with a brief explanation. For a more elaborate description and the precise definition of all the terms used the reader is referred to [4] and [5].

A3: Algorithm 3, authentication algorithm used for authenticating the subscriber

A5: Algorithm 5, cipher algorithm; used for enciphering/deciphering data

A8: Algorithm 8, cipher key generator; used to generate K_c

AuC: Authentication Centre; used to store the keys K_i, A3 and A8 are implemented in the AuC

DECT: Digital European Cordless Telephone

ETSI: European Telecommunications Standards Institute

GSM: Global System for Mobile Communications
Groupe Spécial Mobile (the original name of the body specifying the system within CEPT and ETSI)

HLR: Home Location Register; a register in the HPLMN of the subscriber where (all) information related to the location and the subscription are permanently stored

HPLMN: or Home PLMN; the network with which a subscriber is registered

IMSI: International Mobile Subscriber Identity; the identity which uniquely identifies the subscriber in all GSM networks, used for routing in GSM (not to be confused with the subscriber's mobile telephone number)

K_c: the cipher key; used in A5 to generate the key stream

K_i: the subscriber authentication key; used in A3 and A8

LAI: Location Area Identification; information indicating the location of a cell or a set of cells

ME: Mobile Equipment; the MS without the SIM

MS: Mobile Station; the equipment used to access GSM

PIN: Personal Identification Number; used by the SIM for the verification of the identity of the user

PLMN: Public Land Mobile Network; a network providing communication possibilities for mobile users

PUK: Personal Unblocking Key; used to unblock the GSM application which occurred as a result of three consecutive wrong PIN entries

RAND: a non–predictable number; used as a challenge in the authentication process

SIM: Subscriber Identity Module; the subscriber card containing security and other subscription as well as network related information

SMG: Special Mobile Group; the new name of the ETSI Technical committee formular called Groupe Spécial Mobile

SRES: Signed RESponse; used by the network side to verify the identity of the SIM in the authentication process

TMSI: Temporary Mobile Subscriber Identity; the temporary identity issued by a VLR to provide subscriber identity confidentiality

VLR: Visitor Location Register; the register where the user is (temporarily) registered while in a location controlled by this register

References

1. ANSI X3.92: 1981, *Data Encryption Algorithm.* American National Standards Institute.
2. H.J. Beker and F.C. Piper, *Secure Speech Communications*, Academic Press, London, 1985.
3. C. Brookson, *GSM Security: A Description of the Services*, in: GSM, Digital Cellular Mobile Communications Seminar (ed. F. Hillebrand), Budapest, 1990, 4.5/1–4.5/5.
4. ETSI–GSM, Recommendation GSM 01.02, *General Description of a GSM PLMN*, Version 3.0.0, March 1990.
5. ETSI–GSM, Technical Specification GSM 01.04, *Vocabulary in a GSM PLMN*, Version 3.0.1 (Release 92, Phase 1).
6. ETSI–GSM, Technical Specification GSM 02.09, *Security Aspects*, Version 3.0.1 (Release 92, Phase 1).
7. ETSI–GSM, Technical Specification GSM 02.17, *Subscriber Identity Modules, Functional Characteristics*, Version 3.2.0 (Release 92, Phase 1).
8. ETSI–GSM, Technical Specification GSM 03.02, *Network Architecture*, Version 3.1.4 (Release 92, Phase 1).

9. ETSI–GSM, Technical Specification GSM 03.20, *Security Related Network Functions*, Version 3.3.2 (Release 92, Phase 1).

10. ETSI–GSM, Technical Specification GSM 11.11, *Specifications of the SIM-ME Interface*, Version 3.11.0 (Release 92, Phase 1).

11. ETSI–RES, European Telecommunication Standard, Final Draft prETS 300 175-7, *Digital European Cordless Telecommunications (DECT), Common interface, Part 7: Security features*, May 1992.

12. F. Hillebrand (ed.), *GSM, Digital Cellular Mobile Communications Seminar*, Budapest, 1990.

13. ISO/IEC 7816, *Identification cards–Integrated circuit(s) cards with contacts*.
    Part 1: 1987, *Physical characteristics*.
    Part 2: 1988, *Dimensions and location of the contacts*.
    Part 3: 1989, *Electronic signals and transmission protocols*. (To be amended by T=1.)

14. ISO/IEC 9798, *Information technology–Security techniques–Entity authentication mechanisms*.
    Part 1: 1991, *General model*.
    Part 2: 1992 (CD), *Entity authentication using symmetric techniques*.
    Part 3: 1992 (DIS), *Entity authentication using a public key algorithm*.

15. G. Mazziotto, *Mobile Radio Services. Subscriber Identity Modules*, in: GSM, Digital Cellular Mobile Communications Seminar (ed. F. Hillebrand), Budapest, 1990, 4.4/1–4.4/10.

16. M. Paterson, *Secure Single Chip Microcomputer Manufacture*, in: Smart Card 2000 (ed. D. Chaum), North Holland, Amsterdam, 1991, 29–37.

17. H. van de Pavert, *The Smart Card in the Future European Digital Mobile Telecommunication Network*, in: Smart Card 2000 (ed. D. Chaum), North Holland, Amsterdam, 1991, 85–92.

18. G. Simmons, Contemporary Cryptology, *The Science of Information Integrity*, IEEE Press, Piscataway, NJ, 1992.

19. K. Vedder, *Smart Cards*, Proceedings CompEuro 92, IEEE Computer Society Press, Los Alamitos, 1992, 630–635.

20. M. Walker, *Security in Mobile and Cordless Telecommunications*, Proceedings CompEuro 92, IEEE Computer Society Press, Los Alamitos, 1992, 413–416.

(Local Area) Network Security

Walter Fumy

Siemens AG, Dept. AUT 961

P.O. Box 3220, D-8520 Erlangen, Germany

Abstract

The most obvious threats to information security are those concerning data while being transmitted over a network. A secure network must provide for data confidentiality, for authentication of the originator of a message and for protection against unauthorized changes of the data transmitted. Since local area networks are not confined to small areas anymore, the need for LAN security also has become commonly recognized. Only a few vendors of networking equipment have responded yet to this need, also progress in network security standards is relatively slow.

1 Introduction

A network, in general, is composed of a wide variety of nodes that are connected by transmission media which may be point-to-point or broadcast. Networks supply their users with communication channels. The exchange of messages is controlled by communication protocols. In a typical network, PCs, minicomputers or mainframes have direct access to the network, while dumb terminals are wired into terminal servers that are connected to the network.

The most obvious threats to information security are those concerning data while being transmitted over a network. There is a long standing tradition of methods and techniques for communications security (COMSEC), most of which involve the use of cryptography. In traditional applications of cryptography the main objective used to be data confidentiality. For many commercial network users, however, confidentiality is of secondary importance. A secure network for business applications must rather provide for authentication of the originator of a message and for protection against unauthorized changes of the data transmitted (data integrity).

For a number of years local area networks (LANs) have become very widespread, since they provide a cost effective way of connecting hosts, PCs, terminals, and peripheral devices to each other. LANs used to be confined to small areas and, therefore, did not seem to be much exposed to outsider's attacks. Therefore, security has not been an issue for most of their users. It is, however, important to note that nowadays in many cases LANs extend to much wider areas than the very term indicates. By using repeaters and bridges in order to achieve connectivity between "LAN islands" many installations have grown into very complex topologies covering fairly large areas. This is one of the reasons why the need for LAN security has become more commonly recognized.

Only a few vendors of networking equipment have responded yet to this need by offering concepts and products achieving a high level of security. As progress in network security standards is relatively slow [Kirk 88], many organisations have to implement security measures ahead of those standards.

2 Network Security Issues

In a computer network, one must expect several sorts of malicious activities. Some of them exist with no premeditated intent. Examples of such accidental threats are system malfunctions and software bugs. Major intentional network security threats include:

- **Wire Tapping.** Network media can be tapped. Some media are more, some are less tap resistant. Network media may also radiate information which is a typical weakness of electromagnetic media. It is generally regarded a simple matter to record the data passing through a communications line without detection by the communicating parties. Because in many systems, passwords for access to computers and applications are transmitted over the network in clear, an unprotected communications line is likely to establish a considerably weak spot in an organisation's information security practice.

- **Masquerading.** A masquerade is where an entity pretends to be a different entity. In many networks, nodes can masquerade as other nodes and may then introduce invalid messages into the network that are delivered as if they were genuine.

- **Modification of Messages.** Modification of a message occurs when the content of a data transmission is altered without detection and results in an unauthorized effect.

- **Replay of Messages.** A replay occurs when a message, or part of it, is repeated to produce an unauthorized effect. If it is both possible to record and to introduce messages into a network, nodes may retransmit previously transmitted messages (modified or unmodified).

- **Denial of Service.** Denial of service occurs when a node fails to perform its proper function or acts in a way that prevents other nodes from performing their proper functions. Examples are the suppression of selected messages (e.g. of those directed to a security audit device) or the generation of extra traffic to slow down or disrupt communications.

- **Trapdoors and Trojan Horses.** When an entity is modified to allow an attacker to produce an unauthorized effect on command or at a predetermined event (e.g. a specific date and time), the result is called a trapdoor. A trojan horse also introduces unauthorized functionality into a system. An example for a trojan horse is a relay that copies messages to an unauthorized channel.

Network protocols often rely on the fact that every node is honest. Common mechanisms assume that only the addressed devices will respond to a request for information and that no device will masquerade as another node. In a typical address resolution scenario a station A wants to connect to a network server without knowing its physical address. Therefore, A sends out a broadcast or a multicast message giving its own address and requesting the address of the service required. After having received the physical address from a responding node (which may be the right server or a subverted system) A will try to logon to the service using the received address. Such a protocol obviously enables an attacker to learn user ID and password of A.

For LANs, the easy access to data anywhere along a LAN is one of its biggest security problems. Today's most prominent LANs, Ethernet and Token Ring, as well as the FDDI standard, use either a logical bus or ring access method where all the data messages are sent to everyone on the network. Ring topology networks not only make all data available to every node, but require every station to intercept and forward every frame in order for the LAN to operate properly. A normal station of course is expected to ignore all data not addressed to it.

It is a major benefit of such a broadcast communication technique that new nodes can be added and the network's configuration be changed easily. On the other hand, this transmission technique is inherently insecure since various kinds of unauthorized manipulations to the network are equally simple to implement. Moreover, there is little chance to detect if a node looks at data which is addressed to someone else. Therefore, eavesdropping and masquerading either by misuse of authorized network nodes or by clandestine installation of new ones (e.g. running diagnosis tools) are considered the most important threats to security in LANs.

3   Network Security Architecture

Security measures for communications systems can be realized in various ways. In particular, there are many options for the integration of security services into the communications architecture. A network **Security Architecture** constitutes an overall security blueprint for a network. It describes security services and their interrelationships, and it shows how the security services map onto the network architecture.

In order to extend the field of application of the Basic Reference Model for Open System Interconnection [ISO 7498], ISO (the International Organization for Standardization) has identified a set of security services and possibilities of integrating them into the seven layers of the OSI architecture. The OSI security architecture [ISO 7498-2] supports five primary security services: authentication, access control, data confidentiality, data integrity, and non-repudiation.

- Peer entity authentication is the verification that the communicating entities are the ones claimed (in the context of OSI an entity is a logical or physical endpoint participating in a data exchange).
- Access control provides protection against unauthorized use of resources.
- Data confidentiality provides protection of information from unauthorized disclosure.

- Communications integrity ensures that data is accurately transmitted from source to destination. The network must be able to counter equipment failure as well as actions by persons or processes that are not authorized to alter the data.
- Non-repudiation with proof of origin protects against any attempt by the sender to falsely deny having sent a message, while non-repudiation with proof of delivery protects against any attempt by the recipient to falsely deny having received a message.

Figure 1 shows 14 specific security services identified in ISO 7498-2 and their possible allocation to each layer. It also illustrates that strict adherence to the OSI security architecture would limit LAN specific security services (i.e. security services for layers 1 and 2) to data confidentiality.

Security Service	Layer						
	1	2	3	4	5	6	7
Peer Entity Authentication			X	X			X
Data Origin Authentication			X	X			X
Access Control			X	X			X
Connection Confidentiality	X	X	X	X			X
Connectionless Confidentiality		X	X	X			X
Selective Field Confidentiality							X
Traffic Flow Confidentiality	X		X				X
Connection Integrity with Recovery				X			X
Connection Integrity without Recovery			X	X			X
Selective Field Connection Integrity							X
Connectionless Integrity			X	X			X
Selective Field Connectionless Integrity							X
Non-repudiation, Origin							X
Non-repudiation, Delivery							X

Figure 1: OSI Security Architecture

According to ISO 7498-2, the types of mechanisms that may be used to provide the security services identified are encipherment, digital signature, access control, data integrity, authentication exchange, traffic padding, routing control, and notarization. Figure 2 illustrates which mechanisms, alone or in combintation, are considered to be appropriate for the provision of each type of service.

Security Mechanism	Security Service				
	Authentica-tion	Access Control	Data Confi-dentiality	Data Integrity	Non-Repu-diation
Encipherment	X		X	X	
Digital Signature	X			X	X
Access Control		X			
Data Integrity				X	X
Authentication Exchange	X				
Traffic Padding			X		
Routing Control			X		
Notarization					X

Figure 2: Mapping of Mechanisms and Services

The purpose of **Security Frameworks** is to describe specific functional areas of security (e.g. authentication, access control, confidentiality) and to provide generic solutions to those security problems. The frameworks define the general means for protecting systems or objects and also are concerned with the interactions between systems. Frameworks do not specify the necessary security mechanisms. ISO and CCITT currently develop a number of security frameworks [ISO 10181-x] (see figure 3).

Layer						
1	2	3	4	5	6	7
Authentication Framework						
Access Control Framework						
Integrity Framework						
Non-Repudiation Framework						
Confidentiality Framework						
Security Audit Framework						
Lower Layer Security Model			Upper Layer Security Model			

Figure 3: Security Frameworks and Models

Also under development are upper layers and lower layers **Security Models** each of which elaborates on the OSI security architecture. The intention of the security models is to detail how and where the security elements of the frameworks and other security mechanisms and techniques can be combined to provide various types and levels of security. The upper layers security model covers layers 5 to 7 [ISO 10745], the lower layers security model layers 1 to 4. Figure 3 summarizes the current activities in the areas of OSI security models and frameworks.

4 Network Security Concepts

To achieve a high level of communications security in a network cryptographic mechanisms are to be employed. Encipherment e.g. supports data confidentiality, cryptographic integrity check values (ICVs) can be used to protect data integrity. If the sender-ID field is included· into the calculation of an ICV data origin authentication can additionally be provided. Non-cryptographic security concepts naturally do have some limitations but can be quite efficient, too.

- Maintaining tight controls on access to the network communication media can prevent access by unauthorized nodes. Ideally, access should be restricted to known devices at known access points only. However, physical security measures are hard to maintain in extended environments and cannot detect malicious activities by authorized nodes.

- Cabling strategies can be very important to security. A new trend is the use of structured wiring techniques for LANs. In this case, patch panels are attached to the backbone cabling that provide the link to the floor wiring. Each node has its own cable to the patch panel. An intelligent patch panel can provide some access control by allowing only authorized devices to connect to the network. Another example is the use of fibre media which are relatively difficult to tap into without being detected. The dominant high speed LAN standard in the 1990s is expected to be FDDI (Fibre Distributed Data Interface), a logical ring network.

- Partitioning of a network into physically or logically separate networks for different user groups can prevent attacks from outside nodes. Bridges use address tables to perform frame filtering and forwarding. The address table entries can be the result of a learning process or they can be static and thus provide a measure of access control (managed bridge). To some extend confidentiality can be achieved by confining local traffic to a subnet. Partitioning of a network cannot prevent attacks from inside nodes and it puts limitations on the communication capabilities which may be not practical in many environments.

The use of cryptographic methods for network communications security can prevent wiretapping, masquerading, and modification attacks, and does additionally allow for some access control policy to be implemented [PoKl 79], [VoKe 85]. However, when being embedded into different layers security services can have rather different properties. Specific security concepts for the OSI layers 2, 3 and 4 are discussed in the following section.

5   Security Protocols

There are several goals that should be met by any security protocol. A security protocol has to provide cryptographic security services and should be independent of specific cryptographic algorithms. It should not interfere with the operation of unprotected systems, provide optional communication between protected and unprotected systems, and support transparent operation for protected systems.

The capabilities of a security protocol depend on its location in the OSI reference model. Each entity communicates with entities in the layers above and below across an interface which is realized as one or more Service Access Points (SAPs). Services are provided via the invocation of primitives which specify the function to be performed and are used to pass data and control information. A security protocol placed between layers has access only to the defined primitives and the associated protocol control information while a security protocol integrated in a specific layer has access to all information available within that layer.

Security services provided at any layer of a protocol stack in general protect only the SDU (Service Data Unit) portion of that layer's PDU (Protocol Data Unit). Header information from that layer and from all lower layers is left unprotected. Also, PDUs that originate and terminate at lower layers cannot be protected.

Several specific protocols have been proposed for adding security services to OSI protocols. Most prominent are IEEE SILS (Standard for Interoperable LAN Security) [IEEE 802.10], and the SDNS protocols SP3 (Security Protocol 3) [SDNS SP3] and SP4 (Security Protocol 4) [SDNS SP4]. They are to provide the primary security services confidentiality, integrity, authentication and access control for OSI layers 2, 3 and 4, respectively.

In the OSI architecture, the transport layer is the lowest layer that is strictly end-to-end. From that point of view, for end-to-end security encipherment should either take place at the bottom of the transport layer or at the top of the network layer. The Secure Data Network System (SDNS) project has developed a security architecture within the OSI computer network model which includes security protocols for OSI layers 3 and 4. The SDNS transport protocol, SP4, is defined as an addendum to the ISO transport protocol. The SDNS network protocol, SP3, is defined as a sublayer of the ISO network protocol which resides directly below the transport layer. Both are currently under consideration within ISO (ISO/IEC/JTC1/SC6) for inclusion into the OSI protocol suite [ISO 10736].

The SP3 and SP4 protocols have been developed for consistency with the OSI security architecture and with the ISO protocols. However, due to the similarity of the layer interface between the ISO transport protocol and the connectionless network protocol to the interface between the DoD protocols TCP and IP, SP3 and SP4 implementations can be adapted to work with those DoD protocols as well.

The basic security services provided by SP3 and SP4 are confidentiality (by encrypting the transport PDU) and connectionless integrity (by calculating an integrity check value and appending it to the TPDU). The protocols were designed to be independent of specific encipherment algorithms and the method of key distribution. Because SP3 and SP4 use pairwise keys, data origin authentication is offered in addition. Access control is provided by key management and by security label checking. Both protocols allow various key granularities (for SP4 e.g. a key per end system network SAP pair, a key per end system NSAP pair and security label, or a key per transport connection). Figure 4 shows the generic PDU format for SP3 and SP4.

As mentioned above, security services can have fairly different properties when being embedded into the various layers. A typical example is provided by encrypting data beneath or above the network layer. In the first case routers or packet-switching-nodes must be able to decipher the data in order to evaluate header information. The second option, on the other hand, can provide true end-to-end encryption, because the headers containing routing information are left unencrypted. The impacts on the security properties and the management requirements differ a great deal between the two choices.

Because only the two lowest OSI layers are refered to by the LAN standards, security services for LANs have to be integrated there. Although, this obviously limits the possibility of providing end-to-end security for internet communications, LAN specific solutions do have some advantages:

- There are standardized communication protocols for LANs in layers 1 and 2, whereas many LANs run a variety of protocols in layers 3 and above (e.g. ISO-8073/8473, TCP/IP, or proprietary protocols). By integrating security services into layer 2 a single security system can be used to protect communications between nodes on the network running any kind of higher layer protocol or application.
- For a LAN, layer 2 provides end-to-end communication.
- LANs can be run by central administration. This puts lower requirements on the security management of a LAN specific solution as compared to a more general security system designed for use in an internet environment.

For LANs, the Institute of Electrical and Electronic Engineers (IEEE) is the leading standardization body. A standard protocol for LAN security using cryptographic techniques is currently being developed by its working group 802.10 under the project title *"Standard for Interoperable LAN Security"* (SILS) [IEEE 802.10], [BaKi 89]. SILS is proposing a transparent Secure Data Exchange sublayer (SDE) between the MAC (Media Access Control) and LLC (Logical Link Control) sublayers of the data link layer. Additional protocols for key and network management are under development. The SDE sublayer provides

- data confidentiality (by encrypting the LLC PDU),
- connectionless integrity (by calculating an integrity check value and appending it to the LLC PDU),
- data origin authentication (by placing the sender ID in the protected portion of the security header and/or by key management), and
- access control (by key management).

The SDE sublayer is optional, i.e. not every node in a LAN has to employ the protocol. It depends on external key management for key establishment and for choosing an encipherment algorithm and an integrity algorithm. SILS SDE has the same generic PDU format as the SDNS security protocols (see figure 4).

Clear Header	Protected Header	User Data		ICV
	ICV Calculation			
	Encipherment			

Figure 4: Generic PDU Format for SILS SDE, SP3, and SP4

6   Security Management

For a security protocol to run correctly, various management tasks have to be performed. Particularly, in the case of a cryptographic protocol matching keys have to be available at both ends of the communication channel. The creation and distribution of those keys in a suitable way is one of the primary responsibilities of security management. However, there are other tasks involved in controlling a system as well, e.g. security audit management. Security management standards currently under development [ISO 10164-x] deal with the management of security objects and attributes (e.g. cryptographic keys, access control information) and cover also functions for alarm reporting and for the management of audit trails.

A **Security Audit** is an independent review and examination of system records and activities in order to confirm compliance with the established security policy, to test the adequacy of system controls, and to assist in the detection of security violations and in the analysis of attacks. It requires the recording of security related events (security audit trail). Security audit mechanisms involve comparing the activities of entities against a known profile (e.g. unusual access based on time or location) and analysing security audit data.

The OSI management model introduces the concept of system management and layer management. System management uses all seven layers for monitoring and controlling a network. Layer management acts directly at a single layer. For each protocol at each layer, there is a Layer Manager (LM) associated with that protocol. The main function of a LM is to manage the objects used by a protocol. The operations on the objects are performed by the LM as directed by the System Management Application Entity (SMAE). The communication of LMs and SMAEs is a local matter and defined by the end systems. In figure 5 this internal communication (often refered to as management "cloud") is shown by the inverted L-shaped box.

The objects to be managed are defined with respect to each protocol. For security protocols, some objects will be devoted to supporting the security mechanisms employed. Examples are the cryptographic algorithm applied and its mode of operation, or the keying material involved (cryptographic keys, initialization values etc.). To indicate the protection and separation of security-related objects with respect to other management objects, the concept of a **Security Management Information Base (SMIB)** is introduced. This data base is accessible by the SP as well as by the Key Distribution Protocol (KDP). The SMIB is part of

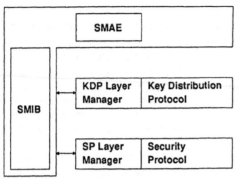

Figure 5

the management "cloud" and provides a communication path from System Management and the KDP to the SP (see figure 5).

When a secure association is set up between a pair of nodes, both systems need to get the keying material necessary for that association. Common security protocols provide mechanisms for identifying the cryptocraphic key in use. The SP4 message format e.g. contains a variable length key identifier field (Key ID) in its clear header, while the SILS SDE message format contains a four byte Security Association Identifier (SAID) field. In addition to the SAID field an optionally supported variable length management defined field is specified that allows the transfer of key management information. To identify keys a common approach is to provide an index into a table containing the keys. However, this forces cryptographic devices to store a potentially large table of keys. One possibility to avoid this storage is to essentially include the (enciphered) association key in those fields. Along with the enciphered association key, the key identifier may consist of a control field that identifies the intended use of the key.

As a result of various design decisions appropriate to different circumstances, key distribution mechanisms vary in many respects. They are based on different types (i.e. symmetric or asymmetric) of cryptographic algorithms, they involve handshakes of different complexity (e.g. two-way, three-way, four-way), and they are used to distribute or generate keying material for different purposes.

Automatic distribution of keys typically employs different data elements to be transmitted. A transaction usually is initiated by requesting a key from some central facility (e.g. a Key Distribution Center), or from the entity a key is to be exchanged with. Information is exchanged between communicating parties for the transmission of keying material or for authentication purposes which may contain keys, or other keying material, such as the distinguished names of entities, initialization vectors, key identities, count or random values.

Even though many key distribution protocols have been designed, little attention has been devoted to the question of how those protocols could be incorporated into the OSI architecture. In [ISO 7498-2] key management is not explicitly confined to any particular layer. In the following, four options for the placement of key distribution protocols within OSI are briefly discussed. A more detailed treatment of this topic can be found in [FuLe 90].

- **Key Distribution Sublayer.** One option is to augment the OSI stack by a KD sublayer which offers service primitives for the establishment and release of a KD connection and for the KD data transfer. Even though such a KD sublayer can be considered independently from the layer implementing the SP, it cannot be decoupled completely. A key for a finer level of granularity (i.e. at an upper layer) cannot be negotiated by a lower layer which implies that the KD-sublayer must not lie below of the SP-layer. On the other hand, the KDP needs to know whether the SP needs a key which implies that the KD-sublayer must not lie above the SP-layer. Therefore, for this alternative the KD protocol has to run in the same layer as the SP.

- **Extend SP to comprise a KD phase.** A connection-oriented SP provides primitives for connection establishment, data transfer and connection release. During the connection establishment phase, the entities that want to communicate establish the parameters necessary for the subsequent phases of the protocol. The second option to incorporate KD is to extend the SP to an XSP such that the keying material becomes part of the parameters negotiated. For this purpose, connection establishment could be divided into two subphases. Decoupling the key exchange phase from the initial non-cryptographic connection establishment is advantageous for two reasons: It provides a higher degree of modularity and a separate sub-phase can be more flexible with respect to the variety of key distribution protocols it can support. The primitives for the data transfer phase and for the connection termination phase of the XSP will coincide with the corresponding primitives of the original SP. For a connectionless SP the extension can be carried out in a similar way.

- **Utilize options of communication protocols.** Many service primitives provide optional parameters (e.g. User-Data). The use of those items is left to bilateral agreements between the communicating parties. One possibility is to utilize the options for key distribution purposes [Rama 90]. Even though this idea sounds quite simple, there are some disadvantages. On the one hand flexibility is limited by the amount of space available in the optional data fields, on the other hand the complexity of the key distribution protocol (e.g. the possible number of handshakes) can be restricted. With TCP the situation is less critical than with ISO protocols. In [Diff 85] it is proposed to establish a common secret key between two TCP-entities by using the options-field of a TCP header. TCP sets up a connection by a three-way handshake. Since the options-field can be arbitrarily long, it can accommodate any handshake information (in [Diff 85] a Diffie-Hellman key exchange protocol [DiHe 76] is suggested).

- **Separate connection for key exchange.** The fourth alternative decouples key distribution from the actual communications association, i.e. a separate connection is established when a key has to be distributed. This can be triggered by a regular system activity or by the layer manager of the SP. The first approach is described in [Herb 88] where the key management system provides keys for link layer protection in LANs. A key is negotiated using a dedicated connection between two nodes. After its establishment the connection is released and a new connection is set up for communications which then can be protected using that key. In the second alternative the layer manager realizes that the SP needs a key and initiates a new connection with a peer KD entity. For obvious reasons this connection cannot be below the layer for which the key is needed. Upon completion of the KDP, the key is written into the SMIB from where it can be accessed by the invoking layer manager (see figure 5). This approach was chosen by the SDNS project [Lamb 88]. There an application layer connection is established that negotiates the key and stores it into the SMIB. It is noteworthy that the SDNS key distribution protocol does not make use of the lower layer security services offered by the SDNS security protocols SP3 or SP4. The SDNS KDP is based on asymmetric cryptographic algorithms. The established association is usually not released after successful key negotiation, but left open throughout the SP-connection. This is due to the fact that key refreshment is required in some modes of SP4 after every 256 or 65536 packets transmitted.

The described alternatives for the placement of a key distribution protocol can be assessed according to several criteria. Main goals are to minimize the development cost, to maximize the efficiency of key management communications, and to comply with existing standards. For any system, the placement of key management functions of course depends on the system architecture, on the capabilities of its components, and on the priorities set for a specific system. If e.g. performance considerations are given top priority alternative 3 will be favorable, whereas if modularity and flexibility are most important one may choose alternative 1 or alternative 4. For a detailed analysis we refer to [FuLe 90].

7 Network Security Devices

So far, only a few vendors of networking equipment have responded to the growing need for network security. There are several constraints on the development of network security devices. Examples are the existence of multiple communication security standards, the need to provide communication security for previously manufactured systems, and last not least economic constraints. Furthermore, export restrictions limit the widespread use of network security devices.

There are three different approaches to the design of a network security device:

- The least costly solution, in general, is to integrate the cryptographic hardware into the end systems. However, if a large number of different systems need to be protected, development costs increase. Additionally, integrated solutions often cannot be used for previously manufactured equipment or for equipment manufactured by other vendors.

- There are several advantages to implementing the security protocols together with the necessary security management functions in self-contained boxes that can be inserted between a node and the network [FuRi 90], [Herb 88]. This approach guarantees independence of other network components and maximum flexibility. Moreover it provides the possibility to use such an in-line box not only to protect a single station but an entire subnet. To this end the security devices must be provided with basic bridge or router capabilities.

- A third possibility is to design a relatively simple outboard cryptographic device that handles only encipherment and decipherment [Herb 90]. This type of security device is totally controlled through frames sent by the node. It recognizes different frame formats passing through and is able to encipher or decipher parts of the frame, or to calculate integrity check functions. In this case the attached end system has to take care of key management and of many details of the communication security protocols.

As an example for the second approach, we will discuss the design philosophy of the *SINEC Kryptobox System* (*SINEC* = Siemens Network Architecture for Automation and Engineering). This outboard cryptographic device has been developed as a cryptographic security product specifically designed to meet the requirements of LANs in a manufacturing environment, based on the IEEE 802.3 LAN standard (Ethernet). Because of the great variety of stations being used there, a concept had to be adopted that did not assume any changes to the hardware and software of existing nodes. Since flexibility was such an important design goal, the system is equally well suited to fit into other environments too, e.g. office communications. The following list summarizes the most relevant requirements to be met by the system design:

- The primary network security services have to be provided, i.e. data confidentiality, data integrity (with or without confidentiality), data origin authentication, and access control. Strong security mechanisms must be used for implementing those services.
- The security system must work independently of network and transport protocols, and of applications.
- The security system must be able to work with the unchanged hardware and software of existing network equipment and nodes and it should be possible for several nodes to share a security device.
- LAN performance and availability should be affected as little as possible and the security system has to be transparent to network users.
- It must be possible to run protected and unprotected traffic on the same network and to install the security system stepwise into an existing network.

From the above requirements the following design decisions can be derived. Cryptographic mechanisms are required to achieve a high level of communications security. Due to performance considerations, for the time being there is little choice but to use a hardware implementation of a fast symmetric encryption scheme, e.g. the well-established DES (Data Encryption Standard) algorithm. By integrating the cryptographic mechanisms into layer 2 the required independence of higher layer protocols is achieved. Although the layer 2 security services data integrity, data origin authentication and access control are not foreseen by ISO 7498-2 (cf. figure 1), it would be unreasonable to exclude them from a LAN security system.

Security management can be implemented on a separate console operating independently of other network management functions. Security management involves the use of a specific protocol running between the management console and each individual security device. In order to protect management communications, the console must be equipped with a security device of its own, which also serves as generator and secure containment for cryptographic keys. The network itself is used for key distribution. Obviously, when being sent over an insecure network the keys must themselves be encrypted. Key-enciphering keys are used much less than the actual communications keys and for that reason need not be replaced as often as those. In this way the introduction of a key hierarchy significantly reduces the number of keys that have to be distributed outside the network.

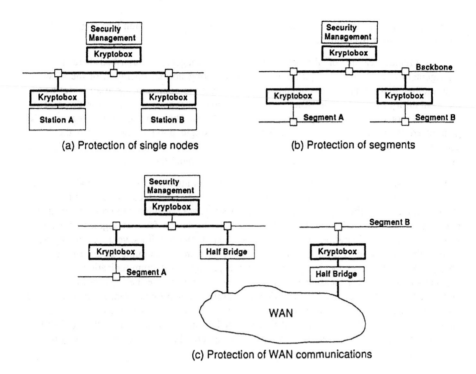

(a) Protection of single nodes   (b) Protection of segments

(c) Protection of WAN communications

Figure 6

In a large network where potentially any two nodes must be able to communicate securely with each other, a great number of keys are required. It would mean a considerable waste of resources, if all of them were generated, protected, distributed, and periodically replaced, even though the majority of them would never actually be used. Consequently in large networks keys have to be generated and distributed on demand. This can only be achieved either by employing rather complex and time consuming key exchange algorithms or by requiring the key management facility being always on-line in order to respond to the key requests of nodes wanting to communicate. However, in the case of a LAN the number of nodes is usually small enough to suggest the adoption of a much simpler scheme; i.e. to create and distribute keys for all cryptographic associations in the network that are both possible and authorized.

It is the security manager's task to decide which security services are to be used with each association. He can also license unprotected communications or prevent certain pairs of nodes from communicating between each other at all. In this way an effective tool of access control for the network is provided. Once the configuration information has been made available to the security management system, it automatically generates the required keys and sends to each security device on the network its specific configuration data and key list.

Of course, functions are included that allow to change the configuration whenever needed or to replace the keys from time to time.

It is a crucial feature of the kind of security management described here that the management console is not required to be on-line during normal operation of the network. This implies that the ability to communicate securely or to communicate at all is not tied to the availability of the console. Another important consequence is that response times of the system are not impaired by key management procedures. Thus, the security management is guaranteed to introduce neither a bottleneck nor a single point of failure into the LAN. Therefore, it can be installed on simple hardware that is not subject to particularly high availability constraints and does not have to be on-line around the clock.

In its simplest mode of operation a *SINEC Kryptobox* is attached to a single node (fig. 6a). Since its functionality includes bridge capabilities it can provide security services to several stations at a time. If there are segments in the LAN that are not subject to security threats the device can be employed as bridge and shared security device for that segment. Figure 6b shows a typical configuration example of a backbone network with several secure segments each of them attached to it through a *SINEC Kryptobox*. In this example the security devices protect data merely on the backbone while the protection of communications inside the segments is left to other measures, such as physical controls. Since the security functions are embedded into layer 2, the *SINEC Kryptobox* system can provide transparent end-to-end security over MAC-bridges. This includes the possibility of secure communications between stations that are attached to distant LANs connected through a wide area network (WAN) by halfbridges (fig. 6c). However, all data must be deciphered before being processed by routers or other gateways operating above layer 2.

References:

[BaKi 89]   Barker, L.K.; Kirkpatrick, K.E.: "The SILS Model for LAN Security", Proceedings of 12th National Computer Security Conference, Baltimore, 1989, 267-276.

[DiHe 76]   Diffie, W.; Hellman, M.E.: "New Directions in Cryptography", IEEE Transactions on Information Theory, 22 (1976), 644-654.

[Diff 85]   Diffie, W.: "Security for the DoD Transmission Control Protocol". Proceedings of Crypto'85, Springer LNCS 218 (1986), 108-127.

[FuLe 90]   Fumy, W.; Leclerc, M.: "Integration of Key Management Protocols into the OSI Architecture", Proceedings of CS'90: Symposium on Computer Security, Fondazione Ugo Bordoni (1991), 151-159.

[FuRi 90]   Fumy, W.; Rieß, H.P.: "Local Area Network Security", Proceedings of CS'90: Symposium on Computer Security, Fondazione Ugo Bordoni (1991), 145-150.

[Herb 88]   Herbison, B.J.: "Developing Ethernet Enhanced-Security System", Proceedings of Crypto'88, Springer LNCS 403 (1990), 507-519.

[Herb 90]   Herbison, B.J.: "Low Cost Outboard Cryptographic Support for SILS and SP4." Proceedings of 13th National Computer Security Conference, Baltimore, 1990, 286-295.

[IEEE 802.10]   IEEE 802.10: *"Standard for Interoperable LAN Security"*, Draft 1/1990.

[ISO 7498] ISO International Standard 7498: *"Open Systems Interconnection: Basic Reference Model"*, 1983.

[ISO 7498-2] ISO International Standard 7498-2: *"Open Systems Interconnection Reference Model - Part 2: Security Architecture"*, 1988.

[ISO 10164-7] ISO/IEC Draft International Standard 10164-7: *Security Management - Security Management Alarm Reporting*, 1990.

[ISO 10164-8] ISO/IEC Committee Draft 10164-8: *Security Management - Security Audit Trail Function*, 1990.

[ISO 10181-2] ISO/IEC Draft International Standard 10181-2: *Security Frameworks for Open Systems - Part 2: Authentication Framework*, 1991.

[ISO 10181-3] ISO/IEC Committee Draft 10181-3: *Security Frameworks for Open Systems - Part 3: Access Control*, 1991.

[ISO 10736] ISO/IEC Committee Draft 10736: *Transport Layer Security Protocol*, 1990.

[ISO 10745] ISO/IEC Committee Draft 10745: *Upper Layers Security Model*, 1991.

[Kirk 88] Kirkpatrick, K.E.: "Standards for Network Security", Proceedings of 11th National Computer Security Conference, Baltimore, 1988, 201-211.

[Lamb 88] Lambert, P.A.: "Architectural Model of the SDNS Key Management Protocol", Proceedings of 11th National Computer Security Conference, Baltimore, 1988, 126-128.

[PoKl 79] Popek, G.J.; Kline, C.S.: "Encryption and Secure Computer Networks", ACM Computing Surveys, 11 (1979), 331-356.

[Rama 90] Ramaswamy, R.: "A Key Management Algorithm for Secure Communications in Open Systems Interconnection Architecture", Computers & Security 9 (1990), 77-84.

[SDNS SP4] Secure Data Network Systems: *"Security protocol 4"*, 7/1988.

[SDNS SP3] Secure Data Network Systems: *"Security protocol 3"*, 2/1989.

[VoKe 85] Voydock, V.L.; Kent, S.T.: "Security in High Level Network Protocols", IEEE Communications Magazine, July 1985, 12-24.

Cryptography Within Phase I
of the EEC-RACE Programme

Antoon Bosselaers, René Govaerts, Joos Vandewalle

ESAT Laboratorium, K.U.Leuven,
K. Mercierlaan 94, B-3001 Heverlee, Belgium

Abstract. In order to pave the way towards commercial use of Integrated Broadband Communications (IBC) in Europe, the Commission of the European Communities has launched the RACE programme. Under this RACE programme pre-competitive and pre-normative work is going on. Most advanced applications in IBC and many services rely on the cost effective provision of integrity mechanisms. Within the first phase of the RACE programme (RACE I) three projects were preoccupied with the provision of these mechanisms in a universal and unified manner to all users of IBC. While project R1025 looked at the overall IBC needs by providing the functional specifications for the global provision of security, projects R1040 and R1047 were providing the necessary technology base.

The goal of the R1040–RIPE consortium was to put forward an ensemble of techniques to meet the anticipated requirements of the future IBC Network in the European Community. A description of the achievements of the RIPE project is given.

1 Integrated Broadband Communications

The European Community has set up a unified European market of about 300 million customers by 1993. In this context, one of the key areas is in the domain of communications and the associated services. As to the lesser developed countries of Eastern Europe, it is widely accepted that efficient telecommunications will be essential both to their economic development and to their relationship with Western Europe. Advanced technologies are considered the most appropriate means of addressing their pressing needs. In view of this market Integrated Broadband Communication (IBC) is planned for commercial use in 1996. The principles of IBC are aligned with the emerging proposals for Broadband ISDN. It will provide high speed channels (64 kbps, 2 Mbps, up to at least 140 Mbps) of image, voice, sound and data communications, and will support a broad spectrum of services, some of which are radically different from those which are familiar today. IBC is the subject of study of the European Economic Community's RACE programme (Research and Development in Advanced Communications Technologies in Europe) [RACE88, RACE90]. RACE covers all aspects of terrestrial networks, satellites and mobile telecommunications. It includes consideration of narrowband networks, distribution networks of all kinds, as well as specific broadband networks, ranging from the provision of broadband, multi-service public networks through Customer Premises Networks to the terminals which will need to be developed to take advantage of the new, high bandwidth services. However RACE is ultimately concerned with services, their definition and their exploitation by end-users. The goal of RACE is to make a major contribution to the objective of the *introduction of Integrated Broadband Communications taking into account the evolving ISDN and national introduction strategies, progressing to Community-wide services by 1995*. Under this RACE programme pre-competitive and pre-normative work is going on. It is clear that the majority of the services offered as well as the management of the network are crucially dependent on the use of cryptographic techniques for their security. Three RACE I projects (R1025 - R1047 - R1040) were preoccupied with the cost effective provision of integrity mechanisms in a *universal* and *unified* manner to all users of IBC.

- Universal Use: Integrity should be available to everybody, that is both private and business users, in an open way without restricting its use to closed-user-groups.
- Unified Use: The way in which integrity is provided needs to be consistent across services and national boundaries.

A RACE Integrity Circle (RINC) ensured the coherence between these three RACE I projects concerned with integrity.

2 Integrity Related RACE Projects

The project R1025 (Functional Specifications of Security and Privacy in IBC) looked at the overall IBC needs by providing the functional specifications for the global provision of security. An integrated functional architecture for IBC security will provide the necessary structured foundation for the development of secure IBC systems. The target objectives of the project were:

- identification of suitable security services and the security management functionality necessary for the protection and separation of IBC and subscriber resources and processes;
- specification of relevant security primitives and attributes necessary for the realization of IBC security;
- specifications for interfaces and protocols.

The aim of the project was to consider security functionality both for the applications of IBC users and for the protection of IBC operation, control and management, and, in addition, to define and specify functions for security management. The work of the project is reflected in six deliverables. The project was discontinued in December 1991.

The other two projects R1047–TIMI (Techniques for Integrity Mechanisms in IBC) and R1040–RIPE (RACE Integrity Primitives Evaluation) were providing the necessary technology base. The former addresses the provision of security mechanisms in a mechanism-independent way, while the latter is especially concerned with the evaluation and provision of the underlying mechanisms. This separation is important in order to maintain the possibility of independent technological evolution.

The main objective of R1047–TIMI was to investigate, specify and demonstrate cost-effective and efficient technical solutions to the problems posed by the provision of universal and unified services and mechanisms for the entire IBC. The more specific objectives were:

- the development of possible integrity concepts in IBC services;
- the evaluation of their cost and performance;
- the development and description of a feasible global integrity system and demonstration of its feasibility for selected applications.

The development of new integrity algorithms was expressively excluded from the scope of this project. A collection of these algorithms and suitable modes of use are provided for by the R1040–RIPE project. Due to internal problems the project R1047 has ended in December 1990.

The RIPE project's main goal was to put forward a comprehensive ensemble of recommended integrity algorithms and suitable modes of use that meet the various security and performance requirements of IBC, for use by R1047 and projects in the pre-normative part of RACE. Rather than developing new algorithms, it was decided to disseminate two successive open calls for algorithms, and to evaluate the algorithms submitted in response to these calls. The project's motivation was the unique opportunity to attain consensus on openly available integrity primitives, which is absolutely required for reasons of interoperability. The members of the RIPE project were: Centre for Mathematics and Computer Science, Amsterdam (prime contractor); Siemens AG; Philips Crypto BV; PTT Research, The Netherlands; Katholieke Universiteit Leuven, and Århus Universitet.

3 The RIPE Calls for Integrity Primitives

3.1 An Open Call for Integrity Primitives

The scope of the RIPE project and the evaluation procedure were fixed after having reached consensus with the main parties involved. The scope includes any digital integrity primitive, except data confidentiality. Figure 1 gives a taxonomy of security primitives. Integrity primitives within the scope of RIPE are shown in boldface, primitives providing data confidentiality, and hence excluded from the scope of RIPE, are shown in italics. In this context it is important to note that in some documents (e.g., [ISO7498]) integrity has a much more restricted meaning.

Early in 1989, a first open call for integrity primitives was disseminated by the RIPE consortium [VdW89]. In response to the first call—which was circulated in a mailing of around 1250 brochures, announced by presentations at the two major conferences on cryptology, Eurocrypt'89 and Crypto'89, and published in the Journal of Cryptology and the IACR Newsletter (International Association for Cryptologic Research)—eighteen submissions were received. Five types of primitives were represented, several well known primitives were submitted, as well as proprietary submissions from major suppliers, thus demonstrating the widespread acceptance and perceived need for the project.

From the eighteen submissions, ten came from academic submitters and eight from industry. The division over different countries was as follows: West Germany 5; U.S.A. 4; Denmark 3; Canada and Japan 2; Belgium and Australia 1. In October 1989, many of the submitters attended special meetings to clarify the submissions.

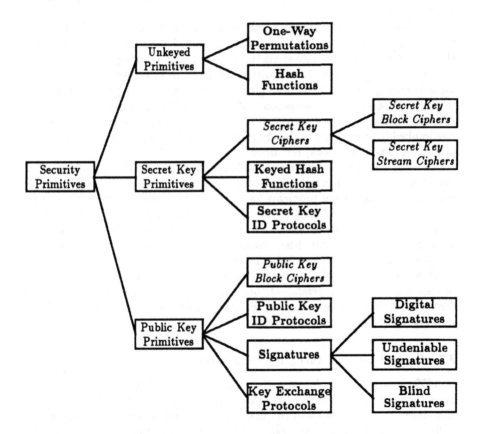

Fig. 1. Taxonomy of security primitives.

In parallel with the work related to obtaining submissions, most of the project resources in the first year have been dedicated to developing a common set of software tools for use in the evaluation phases of both the first and second round. A comprehensive and well-defined set of tools could be specified that was completely operational and documented by the end of the project's first year.

3.2 First Round Evaluation Results

The evaluation phase of the second year was carried out following a fixed procedure. In view of the potential use in IBC the submissions were evaluated with respect to three aspects: functionality, modes of use, and performance. The evaluation comprised computer simulation, statistical verification, and analysis of mathematical structures, particularly to verify the integrity properties. Because of the limited resources and time period, it was decided that if any flaw was identified, the submitter would not be allowed to patch the flaw, thus preventing a moving target.

The first round submissions belonged to one or more of the following five types of integrity primitives: hash functions, keyed hash functions, public key identification protocols, digital signatures, and key exchange protocols. The reader is referred to Fig. 2 for a distribution of the submitted primitives over these five categories. Note that a single submission can contain primitives in more than one category. Five submissions could be rejected in a preliminary screening. After the main phase of the evaluation and after taking into account deficiencies implied by work done in the cryptographic community (external work), ten primitives (six submissions) remained.

These remaining submissions showed significant potential, but each required modification and/or further specification by the submitters. Four of these six showed minor functional problems, one was incompletely specified and one required more analysis. In most of the cases, it was clear how the problems could be avoided. It was however decided to stick to the agreed policy, that is, to inform the submitters about these results and to ask them for a resubmission of modified and/or further specified primitives.

However, the first call for integrity primitives and the subsequent evaluation process was certainly successful. On one hand, important flaws were identified in several submitted schemes, and on the other hand a selection of ten primitives showing significant potential survived.

3.3 Second Call for Integrity Primitives

In 1989, it was already foreseen that some first round submissions would require fixing of functional problems. Moreover, the period of 9 months between announcement of the call and the deadline for submission was relatively short. A final argument for a second call is that work on functional specifications for security within RACE had started only in 1989.

So, early in 1991, a second open call for integrity primitives was disseminated within the cryptographic community [Pre91]. It was circulated in essentially the same ways as the first call. Additionally, formal invitations were made to a selected number of potential submitters, in order to improve coverage of the set of integrity primitives in the taxonomy. In response to the second call thirteen submissions were received, five of them could be considered improved versions of submissions from the first call. Added to two submissions retained from the first round, this resulted in fifteen second round submissions.

Of the fifteen submissions, ten came from academic submitters and the other five from industry. The division over different countries was as follows: Germany 6; U.S.A. 3; Belgium and the Netherlands 2; Canada and Denmark 1. In June 1991 meetings with submitters were again held.

Up until the second call's submission deadline the common set of software tools developed in the first round was upgraded and adjusted, i.e., the set was generalized, extended, and otherwise improved to meet new requirements that emerged. In addition, a number of new (advanced) tools was developed and a pre-evaluation was made of submissions that were expected or retained from the first round.

Primitives	Hash Functions	Keyed Hash Functions	Secret Key ID-Protocols	Public key ID-Protocols	Digital Signatures	Key Exchange Protocols
received round 1	14	8	2	4	3	3
remaining round 1	2	0	0	3	2	3
received round 2	7	2	0	4	2	4
recommended	2	2	1	1	1	2

Fig. 2. Categories of RIPE submissions.

3.4 Second Round Evaluation Results

In order to assure that the recommended integrity primitives result in a comprehensive coverage of IBC requirements, the following sources for primitives were taken into account in the second evaluation phase: the responses to the second call, the revised versions of the first round primitives believed to be promising after the first evaluation, and other primitives proposed in open literature and by the international standards community. Additionally, as mentioned before, some submissions were invited. Essentially the same evaluation techniques were used as for the first call.

The second round submissions belonged to the same five types of integrity primitives as the first round submissions. Five submissions could be rejected in a preliminary screening. After the main phase of the evaluation six primitives (four submissions) remained. Three more primitives were adapted from the literature for more comprehensive coverage (see Fig. 2 for more details).

The recommended primitives recommended are described in the final report of the project. This report includes: a survey of the kind of problems generally encountered in information security and the particular kinds of solutions offered by integrity primitives; a list of the selected algorithms; precise specifications for these algorithms and their recommended modes of use; evaluation of their security; performance estimates of software and hardware implementations; software implementation guidelines and test values. The report contains also two useful appendices, one which serves as a guide to RSA key generation, while the other introduces the basic principles of the modular arithmetic computations needed. The report is currently available from CWI (P.O.B. 4079, 1009 AB Amsterdam, The Netherlands) as CWI Report CS-R9324. The following primitives are described in this report:

Hash Functions: MDC-4 and RIPEMD. MDC-4 (MDC stands for Manipulation Detection Code), invented by D. Coppersmith, S. Pilpel, C.H. Meyer, S.M. Matyas, M.M. Hyden, J. Oseas, B. Brachtl, and M. Schilling is a hash function based on a symmetric block cipher and is recommended for use with a DES-engine. RIPEMD (MD in this case stands for Message Digest) is based on the publicly known hash function MD4, which was submitted to RIPE by RSA Inc. However, due to partial attacks on the first two out of three rounds (found by R. Merkle) and also on the last two rounds (discovered by the RIPE-team), it was concluded that applying the design principle of MDC-4 to MD4 would make it a sufficiently secure unkeyed hash function, which would be very fast in software.

Keyed Hash Functions: RIPEMAC and IBC-Hash. The RIPEMAC algorithm (MAC stands for Message Authentication Code) is inspired by the standardized "CBC-MAC" function (see, e.g., ISO/IEC 9797). Like CBC-MAC, RIPEMAC is based on the block chaining mode of a block cipher. It has the additional feature that it cannot be directly used for concealment. Two variants of RIPEMAC are given, which are based on single and triple encryption— corresponding to increasing levels of security. The second keyed hash function, IBC-hash, has been designed by members of the RIPE-team, David Chaum, Maarten van der Ham and Bert den Boer. IBC-Hash has the distinguished feature of being both provably secure and rather fast in both hardware and software.

Secret Key ID-Protocols: SKID. The primitive named SKID actually contains two secret key identification protocols that can be used to provide entity authentication. The two variants are for unidirectional and bidirectional authentication. Both are based on a keyed one-way function rather than on a block cipher. Their design emphasizes the integrity oriented approach of the project.

Public-Key Primitives: RSA and COMSET. The RSA public key scheme was submitted to RIPE by RSA Inc. The digital signature based on ISO/IEC

9796 is thus one recommended mode of use for RSA, where the input typically would be the result of one of the RIPE unkeyed integrity primitives applied to a message. Another recommended mode of use provides key forwarding. COMSET is a public key identification protocol which has been designed and submitted by a research group at Aarhus University, Jørgen Brandt, Ivan Damgård, Peter Landrock and Torben Pedersen. COMSET is a zero-knowledge mutual identification scheme, which in addition allows for secret key exchange.

4 Conclusion

The RACE programme, focused on IBC and the development of technology for introduction of commercial IBC services in 1995, has been introduced. A short overview of the aims and work of three security related RACE I projects has been presented. Finally, a detailed account of the objectives, the work and the achievements of the RIPE project has been given. The project has carried out its planned acquisition and evaluation of integrity primitives. It has been able to put forward a comprehensive yet carefully evaluated and specified portfolio of integrity primitives.

References

[ISO7498] ISO/IEC 7498-2, *"Information processing – Open systems interconnection – Basic reference model – Part 2: Security architecture,"* ISO/IEC, 1987.

[Pre91] B. Preneel, D. Chaum, W. Fumy, C.J.A. Jansen, P. Landrock, and G. Roelofsen, "Race Integrity Primitives Evaluation (RIPE): A status report," *Advances in Cryptology, Proc. Eurocrypt'91, LNCS 547*, D.W. Davies, Ed., Springer-Verlag, 1991, pp. 547–551.

[VdW89] J. Vandewalle, D. Chaum, W. Fumy, C.J.A. Jansen, P. Landrock, and G. Roelofsen, "A European call for cryptographic algorithms: RIPE; Race Integrity Primitives Evaluation," *Advances in Cryptology, Proc. Eurocrypt'88, LNCS 330*, C.G. Günther, Ed., Springer-Verlag, 1988, pp. 267–271.

[RACE88] *"RACE Workplan '89,"* Commission of the European Communities, 1988, Rue de la Loi 200, B-1049, Brussels, Belgium.

[RACE90] *"RACE Workplan '91,"* Commission of the European Communities, 1990, Rue de la Loi 200, B-1049, Brussels, Belgium.

EDI Security

Gordon Lennox

Directorate General XIII :
Telecommunications, Information Industries and Innovation.
Commission of the European Communities

Abstract. *Gordon Lennox is responsible for security activities in the TEDIS programme, a Community programme dealing with systems for the electronic transfer of trade data. He has also been involved in other EC projects: projects which dealt with the confidentiality and security of data, the protection of computer programmes, and the information society and its environment.*

Please note that the text reflects the author's own views and does not commit the European Commission in any way.

1 Introduction

The use of EDI and the Single European Market will each have a great impact on the ways in which companies operating in the EC conduct their business.

The need has been identified to look at the security requirements for EDI systems to be used within the EC. A Community programme, TEDIS (Trade EDI Systems), was established to address this issue, among others.

This text identifies some of the security issues facing the use of EDI in Europe and mentions some of the activities under way to meet the challenge.

2 EDI

EDI (Electronic Data Interchange) is the electronic transfer from computer to computer of commercial or administrative data using an agreed standard to structure the message data.

The set of international EDI standards is referred to as UN/EDIFACT (United Nations / EDI for Administration, Commerce and Transport) [1]. ISO 9735 gives the EDIFACT syntax. ISO 7372 is the Trade Data Element Directory. UNSMs refers to UN standard messages such as invoice or purchase order.

EDI can be carried out in many ways. Data may be exchanged directly between companies or there may be one or more third-party service suppliers involved. The transmission mechanisms can be file-based or message-based. X.400

is often cited as the vehicle of choice for EDI in the future but in certain sectors and regions there is a tendency to use FTAM-like solutions – for example the French banks' ETEBAC-5 system.

EDI forms an important part of what can be referred to as Computer-Aided Trade (CAT) or Electronic Commerce, terms which cover many networked-based applications – tele-shopping, tele-banking, direct or manual entry, EFT, E-POS, directory services etc..

The term EDI does not normally refer to interactive telematic services such as tele-banking or tele-shopping. However consideration is being given to QR-EDI (quick reaction or query-response EDI) where there may be an exchange of several short messages linked with one single business transaction and during a single session.

3 Information Security

Information security is about maintaining access to information while protecting it from accidental destruction, unauthorized modification or unwanted disclosure. Or in a slightly more abbreviated form: protecting the availability, integrity and confidentiality of information.

In the past security was often seen as being mainly to do with preserving the confidentiality of information. While the limits of confidentiality as an aim are being tested with systems which permit anonymous use, in recent years there has been an increasing tendency to emphasise the availability and integrity aspects.

The move to EDI and associated techniques such as just-in-time stock management (or even "JIT administration") brings many benefits. At the same time there can be an increased level of vulnerability. There may be less capital tied-up in stock but if the network or system goes down then shelves can be left empty or production lines may come to a halt. These are to an extent the normal problems associated with increased computerisation. Nevertheless, they deserve to be re-examined in the light of how the use of EDI can change a business.

By removing the limits imposed by the flow of paper EDI changes the impact of some of the basic constraints on a business. Other factors which may be brought to the fore include quality variability, order sizing, transport reliability. Significant and successful implementation of EDI implies a real understanding, and if necessary a re-appraisal, of the fundamentals of the business.

With EDI transactions are carried out more quickly and the business cycle is completed more quickly with very much less human intervention. Priority needs to be given to prevention of errors and problems rather than correction. Integrity and local transparency become essential system attributes.

While we should not underestimate the wider scope of EDI security, many of the problems can be tackled by the individual company. However the emphasis in much of the discussion and recent work on EDI security has been on the inter-company aspects.

EDI systems are multi-owner systems. No one person, no one organisation will control the full system used in an EDI exchange; the two (or more) end-systems, the network, the network services: all have to be taken into account. Responsibilities need to be assigned.

Business relationships may be very asymmetric and yet both partners still need reasonable and equitable levels of security. The old saying about being "as strong as the weakest link in the chain" needs to be turned around. Systems need to be designed and responsibilities assigned so that the "weak link" does not determine the strength of the overall system and does not present a global threat.

EDI security is perhaps more about providing proof, for auditors and lawyers among others, about what happened than it is about protecting the system against malicious acts: identifying the known individual and linking him with a message rather than protecting against the unknown intruder. This "proof" needs to be relatively cheap to produce and easy to understand. It should be such that it discourages fraud and encourages trust.

Security in EDI should provide enough trust for business to be carried out in a relationship which has a normal degree of mutual suspicion. When the buying department has a new supplier or the sales department has a new client, it should be possible to start trading using EDI quickly without being constrained by concerns about lack of security.

Security in EDI is about using using security tools and techniques to help structure and manage the business relationship in an electronic environment.

4   TEDIS

The TEDIS (Trade EDI Systems) programme was established, initially within the European Community and since opened to EFTA, to address matters relating to the use of EDI across Europe. The programme had a duration of two years and started at the beginning of 1988.

The basis for the programme (1988-1989) can be found in two documents[1]:

COM(86) 662 final, 1 December 1986. Communication from the Commission on Trade Electronic Data Interchange Systems (TEDIS), Proposal for a Council Regulation introducing the preparatory phase of a Community programme on trade electronic data interchange systems.

---

[1] COM documents can be obtained from the Office for Official Publications of the European Communities; L-2985 Luxembourg.

Council decision of 5 October 1987 introducing a communications network Community programme on trade electronic data interchange systems (TEDIS). O.J. No L 285/35, 8.10.87.

TEDIS was mainly about increasing awareness of EDI and the associated standards and about coordinating the various activities underway. There are two types of coordination group: sector groups such as ODETTE, CEFIC and EDIFICE and interest groups around subjects such as legal aspects, telecommunications and security.

A report on the activities of the programme was published:

COM(90) 361 final, 25 July 1990. TEDIS programme 1988-1989 Activity Report.

A proposal for a second phase has been prepared:

COM(90) 475 final, 7 November 1990. Commission Communication on electronic data interchange (EDI) using telecommunications services networks. Proposal for a Council Decision establishing the second phase of the TEDIS programme (Trade Electronic Data Interchange Systems).

5 Security Activities

Consideration has been given for some time to the needs for logical security in open systems. There is a standard OSI security architecture – ISO 7498-2 (CCITT X.800) – which provides a basic vocabulary and framework. There exists a group of organisations in various EC and EFTA states, the TeleTrust group, which has been promoting some of the concepts involved in trading electronically.

Probably the most significant of the recent activities was the preparation of the report on "Security in Open Networks" [2] produced for SOGITS (Senior Officials Group – IT Standardization).

In preparing this report EDI was used as the model application and in the survey involved the most highly rated facilities were those related to:

- User authentication;
- Message integrity,
- Network service availability;
- Confirmation of delivery;
- User authority to use network services;
- Message privacy / confidentiality;
- Network service – level of operation security;
- Auditability;
- Security elements in EDI message structures.

In addition the legal aspects were also considered important: a uniform (EC) basis for contracts; legally enforceable, uniform codes of conduct; harmonised rules of evidence. CEN-CENELEC organised a workshop [3] based on the SOG-ITS report and significant work is continuing there [4].

During 1989 a workshop [9] was organised by TEDIS to look at some of the requirements, particularly those related to non-repudiation, in more detail.

The workshop involved people from various groups (EDI users and research workers; people from network and service suppliers, the PTTs, computer manufacturers, software houses, security equipment suppliers, and the standards area) and they looked at the basic requirements from four points of view: users, standard makers, systems support and service suppliers.

The basic conclusions from the workshop were that the technology is largely available, there is scope for new services and products and while no single solution may be appropriate some structuring was needed.

The use of digital signatures[2] attached directly to EDI messages attracted a lot of attention. It was proposed that something similar to the ANSI approach as found in X12.42 [6] and X12.58 [7] be adopted. That is putting security segments in the EDIFACT-based interchange but using digital signatures as described in X.509 [8].

TEDIS launched a call for proposals in September 1989 [9] and as a result a small series of security projects have been funded: to provide the tools for increasing awareness of EDI security (a handbook and a series of factsheets) and of digital signatures (a PC-based demonstration), to look at the service infrastructure required (notaries, archives, issuers of key certificates, etc.) and to examine how digital signatures could be used in EDIFACT messages.

The report on how digital signatures, "A proposal concerning the use of Digital Signatures in EDIFACT" prepared by Cryptomathic A/S was published in November 1990.

In addition to the TEDIS work a special team was set up at the beginning of 1990 within the EDIFACT banking message development group, MD4B, to look at security. This group published a report: "A Security Framework for EDIFACT".

Both reports provide possible solutions to perceived needs in EDI security. The common factors include:

- no use is made of lower ISO/OSI layers; 'end to end' solutions are provided (EDIFACT is independent of the underlying transport mechanism);
- the approach at the message level is to insert new segments after UNH and before UNT.

---

[2] Data appended to, or cryptographic transformation of, a data unit that allows a recipient of a data unit to prove the source and integrity of the data unit and to protect against forgery (e.g. by the recipient).

Work is proceeding on security related to EDI in both the EDIFACT and X.400 areas but more and more interest is being directed towards X.500 [10]. Two specific areas are worthy of mention: the work in CCITT on X.435 [11] and the ISO work on the conceptual model [12].

Using EDI in an open and secure fashion implies a need for a security infrastructure:

- trusted key issuers;
- key-management facilities;
- user registration;
- notary services;
- archiving;
- security gateways.

Such services – logical security services – are being proposed in Germany and in France. The development of these services and other will be taken into consideration during the second phase of TEDIS.

6 Legal Aspects

Since the beginning there has been a clear link between the security and the legal aspects of the TEDIS programme. The legal aspects were considered important in the SOGITS report referred to above and certain requirements were identified: a uniform (EC) basis for contracts; legally enforceable, uniform codes of conduct; harmonised rules of evidence.

The legal part of the TEDIS programme has involved examining potential barriers to EDI arising from legislation, ensuring coordination between the legal working parties of EDI groups and studying future action necessary by the Commission.

A report on "The legal situation of the Member States with regards EDI" has been produced. This report contains an analysis of the relevant legislation in each of the twelve Member States. It deals with the legal requirements and constraints with regards the transmission of trade documents. The report is being extended to cover the EFTA countries.

Three main types of constraints were identified: the need to prepare, deliver and store paper-based documents with hand-written signatures, obstacles linked to requirements for proof and difficulties in determining the time and place of the conclusion of transactions involving EDI.

The authors of the report concluded that action should be undertaken to adapt accounting and fiscal law and the law on evidence.

The TEDIS Legal workshop in June 1989 examined: contract formation; responsibility and liability in an EDI environment; impact of data protection (privacy) legislation on EDI; the electronic notary.

The results of the workshop demonstrated a need for further investigation and projects are now under way dealing with: contract formation; liability of network operators; trusted third parties and similar services.

Several legal working groups have been preparing interchange agreements, based often on the UNCID rules (the ICC code of conduct for interchange of trade data by teletransmission). A single model interchange agreement is now being produced with the help of lawyers from interested groups.

While it is certainly possible to deal with the legal difficulties, the current legal situation is not really satisfactory. The recommendations and resolutions of international bodies such as the Council of Europe, UNCITRAL and the Customs Cooperation Council indicate that concrete action may be required.

7   Management Issues

It is often said that EDI is not a technical issue but a business issue. This is sometimes taken to mean that the technical issues can be ignored or put to one side. Successful implementation of EDI however requires an understanding of the technology so that it can be best exploited to meet business needs.

EDI is most successful when it is business-driven / user-driven. EDI security has to take into account the companies' trading risk (financial exposure, contractual liabilities). The limits to action however are often the users' understanding of the technology and the users' awareness of security. EDI security is not only a concern of the DP department. All the key players, including the companies' lawyers and the companies' auditors, need to understand and appreciate the move to EDI.

This understanding needs to cover the implementation / exploitation process and how this changes the perception of security. Initial EDI experiences tend to be limited to well-established trading relationships where there is a degree of tolerance and trust. The use of paper documents is prone to error and one of the immediate bonuses of EDI is that the error rate drops dramatically. So there is often no perceived need for security at that early stage. Security becomes important as EDI usage starts to mature. As the business cycle begins to speed up then system availability and preventive measures become important. As the company seeks to extend its usage of EDI to new suppliers and customers then providing proof of responsibility becomes more important.

Management should keep this path in mind. They need to think not only about where they are in terms of using EDI but also where they are going.

This new environment should not lead people to forgetting the basics of good business practice. The separation of roles and functions as a means of ensuring security is still important with all aspects of EDI, and this is increasingly seen to be so by key users. For example, the EDI service supplier can accept responsibility for levels of service availability and global data integrity while another supplier deals with tools for user authentication, etc..

In recent years there has been a change in perception of security in some companies. With a proper understanding of the possibilities afforded by secure applications and systems, security need no longer be viewed as something negative: a cost justified in terms of what might go wrong. Security can be an enabling factor. A secure application can provide the platform for improved client-supplier relationships and for new services. A secure EDI implementation can help make a company more responsive to a changing environment.

Still other companies have seen that there is scope for new third-party services. These can be on-line network services, used either on a transaction basis or on an ad hoc basis such as directories or notaries, or they can auxiliary services such as the provision of certificates or signature devices.

8 Conclusions

In multi-owner systems security becomes negotiable. It has to be standardised to an extent. The security standards – procedures and mechanisms – have to be applicable in a wide set of environments. They have to be acceptable to many people and usable by many people.

No single solution has been identified to meet all the basic requirements. Perhaps no single solution is desirable: choice can be an important element in establishing trust. At the same time there is a need and a possibility for some structure to make the decisions easier and implementation simpler.

Work is proceeding on security related to EDI in both the EDIFACT and X.400 areas but more and more interest is being directed towards X.500.

EDI security can involve on-line and off-line services. The network provider may not always be the most appropriate supplier of certain security related services. Other players, such as banks and credit card companies, will probably be involved.

EDI security activities must meet the real business, legal and social needs, as reflected in the underlying contracts and the relevant legislation. The immediate aim is that an EDI message be acceptable as proof of a transaction by a third-party, that EDI be legally credible. The overall goal is an open and secure environment for computer-aided trade.

References

1. *"Introduction to UN/EDIFACT,"* published by the UN/EDIFACT Rapporteurs teams.
2. EEC report *"Security in Open Networks,"* SOGITS working document Nr 303, January 1989.
3. CEN/CENELEC workshop proceedings. *"Security aspects in OSI functional standards,"* Brussels, October 1989.
4. CEN/CENELEC Ad Hoc Group on Security Standards – *"Towards a taxonomy for standardisation of security,"* E.J. Humphreys, March 1990.
5. *"The TEDIS – EDI Security Workshop Report,"*, Brussels, 1989. See also TEDIS legal workshop report.
6. X12.42, *"Cryptographic service message transaction set."*
7. X12.58, *"Electronic data interchange security structures."*
8. X.509, *"The Directory – Authentication Framework."*
9. Official Journal C230 and S271 of 07.09.1989.
10. *"Security in directories,"* John Draper, The Electronic Directories Conference, London, 1990.
11. CCITT Draft Recommendation X.435, *"Message Handling Systems: EDI Messaging System,"* CCITT Draft Recommendation X.435. Message Handling: EDI Messaging Service.
12. ISO/IEC JTC1/SWG-EDI N 177. Report on the conceptual model for EDI (*"Open EDI Model"*).

Postscript

Since this text was written for the 1991 ESAT course there has been significant progress in the area of EDI security. There is now an established EDIFACT Joint Security Group and a Western European SIG. There has also been progress in the standardisation of signatures. The foundations are being laid for a secure electronic marketplace.

GL. October 1991, Brussels.

AXYTRANS:

physical funds transport and DES [1]

Marc Geoffroy
AXYTRANS,
24 Rue de la Redoute,
B.P. N°10, Z.I.-St.Apollinaire, F-21019 Dijon Cedex

Ronny Bjones and Hedwig Cnudde
CRYPTECH,
Leuvensesteenweg 510-512 Bus 18, B-1930 Zaventem

Abstract

Currently, the transport of valuables and their storage is based upon physical determent. Very often, this leads to the escalation of violence. Therefore, AXYTRANS developed a whole new way of thinking with regard to the transport of valuables (i.c. paper money, bank cards, ...): AXYTRANS. This system aims to degrade the valuables in the case of an agression: the value of the original bank note or bank card is reduced to that of a piece of paper or plastic. The practical implementation of the system required the practical knowledge of a variety of different disciplines: basic electronics, physical and mechanical security, cryptography, computer security, ... In this project, CRYPTECH took care of all the logical security aspects and of the implementation of the *centre serveur* [2]. The system is currently being tested by the PTT in France and in Belgium.

1 General description of the system

Currently, the transport of valuables and their storage is based upon physical determent. Very often, this leads to the escalation of violence.

Therefore, AXYTRANS developed a whole new way of thinking with regard to the transport

---

[1] This paper has previously been presented at SECURICOM 90, CNIT-Paris La Défense, France.

[2] The first concepts of the AXYTRANS system have been written in French; to avoid distortion by traduction into English, some expressions have been kept in French but they are printed in italics.

of valuables (paper money, bank cards, ...). This system aims to degrade the valuables in the case of an agression: the value of the original bank note or bank card is reduced to that of a piece of paper or plastic.

As a result it is no longer necessary to arm the *convoyeurs* and to armour the *véhicules*.

1.1 General principles

It is easy to understand that the transport of valuables and their storage cause a lot of danger due to their high monetary value. If now these bank notes or bank cards could be considered as pieces of paper and plastic, they would become totally worthless for aggressors. The AXYTRANS system indeed will degrade a bank note or a bank card to an ordinary piece of paper or plastic in the case of an aggression.

The order to degrade the valuables is not given by a human being but by some piece of electronic. Obviously, the system has to be able to detect the external events that point to an aggression.

At any time, the boxes containing the valuables (the *conteneurs*) are connected to some component of the system; this connection can be physical or logical. This link is really necessary because the *conteneurs* must know their status and position at each time.

From the first moment the system suspects that something is going wrong, the valuables in the *conteneurs* are degraded.

1.2 Components of the AXYTRANS system

The system can be divided into 3 kinds of components:

- mobile components (*conteneur* and *véhicule*);
- components that constitute the infrastructure (*station*);
- the decision taking component (*centre serveur*).

1.2.1 The conteneur

The system that executes the degradation

This consists out of:

- 2 small containers filled with unwashable paint which marks the valuables by colouring them (making them worthless);
- 1 ignition mechanism and 2 detonators, which have a double function:
  - separating all bank notes or bank cards;
  - tearing open the small containers so that the paint flows over all valuables;
- 1 electronic circuit for the activation of the ignition mechanism; as a matter of fact, this electronic circuit holds up the ignition as long as it receives a certain order from some control circuit.

The money tray

This money tray consists of:

- the box;
- the drawer.

The internal side is covered with 2 aluminium coats wich are separated by an isolating harsh coat, constituting a flat capacitor; this capacitor is connected to the control circuit.

Any attempt of penetration or intrusion will change the capacitance allerting the control circuit, that will allert the electronic circuit for the activation of the ignition mechanism ...

The electronic control circuit

This is an electronic control card constituting the intelligence of the *conteneur*. It detects each form of agression. Besides that, it is able to communicate with the *station* and with the *centre serveur*.

This control card is attached to a physical connector at the back side of the *conteneur*; this connector allows the *conteneur* to be pushed into a receptacle of the *station*.

1.2.2 The station

User interface

It is the basis for the exploitation of the system: it allows authorized people to communicate with the *conteneur* or with another *station* through the *centre serveur*.

This man-machine communication happens through the keyboard and display of a *minitel* terminal; by means of menus, the authorized people are allowed to select the following functions:

- The demand for valuables: this is a service allowing a bank office to ask for the delivery of one or more *conteneurs* by a *convoyeur*.
- The controlled unlocking and opening of a *conteneur*.
- The expedition of valuables: this procedure allows the authorized people to program a *conteneur* with a specific destination and to load the *conteneur* with valuables.
- The collection of the *conteneurs*: this procedure allows a *convoyeur* to collect those *conteneurs* he is responsible for.
- The demand for information: e.g. to know the destination and contents of a *conteneur*.

Communication between *station* and *centre serveur*

The *station* is able to communicate with the *centre serveur* through the public X.25 network; the *station* is connected to the nearest PAD (Packet Assembler/Disassembler) trough the PSTN (Public Switched Telephone Network).

Communication between *station* and *conteneur*

This is a local network that allows a *station* to communicate with up to 50 *conteneurs* and to control these.

Identification - cfr. section 2 on "Security aspects of AXYTRANS"

Using cryptography, the *station* is able to authenticate itself to a *conteneur* (through the *centre serveur*). This guarantees to the *conteneur* that it has arrived at the correct *station*.

Access card - cfr. section 2 on "Security aspects of AXYTRANS"

This is small hand-held device (looking like a pocket calculator) that is used for the authentication of people.

1.2.3 The véhicule

Since the *conteneur* alone is sufficient for the integrity of the transported valuables, an armoured car is not needed anymore.

Nevertheless to facilitate exploitation, the *véhicule* will be equipped with receptacles (to hold the *conteneurs* during the transport), with some controlling electronic and with a terminal (style *minitel*) to identify the *conteneurs*.

1.2.4 The centre serveur

This is the decision taking component of the AXYTRANS system; it can be considered as the heart of the system and it is located in a safe and secret place.

The *centre serveur* has various functions:

- the global control of the system (it authorizes the opening or the collection of a *conteneur*, it manages and controls a whole series of transportation parameters);
- the cryptographic key management (generation and distribution of those keys);
- database management (maintenance and consulting of basic information);
- administrative exploitation (statistics, accounting, invoicing, ...).

The following 3 characteristics are extremely important for this *centre serveur* and its mainframe:

- Reliability: because the *centre serveur* is the decision taking component of the AXYTRANS system, it is of vital importance that it is 100% operational during the whole period. Not one breakdown (even not for preventive maintenance) is allowed. That is the reason why a Fault Tolerant OLTP (On-Line Transaction Processing) machine has been chosen.
- Performance: the configuration of the *centre serveur* should dispose of sufficient computing power to guarantee the service.
- Security: this aspected is treated in much more detail in the next section.

2 Security aspects of AXYTRANS

The main objective of the AXYTRANS system is the <u>secure</u> transport of *conteneurs* from one *site* to another. This transport is supervised by the *centre serveur*.

This section describes the various aspects that make up the global security of the AXYTRANS system, i.e.:

- the logical protection of the transport of the *conteneurs*;
- the logical protection of the *centre serveur*;
- the key management.

2.1 Logical protection of the transport of the conteneurs

As stated above the main objective of the AXYTRANS system is the secure transport of *conteneurs* supervised by the *centre serveur*; the way how the *conteneurs* are physically moved is not relevant for the *centre serveur*.

When a *conteneur* leaves the *site*, it is removed from the *station* where it was kept. When it arrives into a *site*, it is entered into the local *station*.

Such a transport is schematically depicted in the figure below.

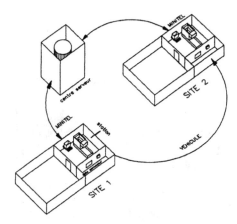

Note that the function of the *stations* is twofold:

- to physically hold the *conteneurs*;
- to support the communication between the *conteneurs* and the *centre serveur*.

2.1.1 Security needs

Let us first identify the parties that are involved at the various moments of a transport cycle:

- the *conteneur*;
- the *centre serveur*;
- the *station*;
- the *responsable d'agence*;
- the *convoyeur*.

All these parties do interact with each other during such a transport cycle: e.g. at a certain moment the *convoyeur* goes to the *site* to take the *conteneurs* with him that have to be transported to another *site* and this transport will be supervised by the *centre serveur*.

It is obvious that all these parties may be a potential threat to the security of the AXYTRANS system; so they may all distrust each other. Therefore, all parties need mechanisms to authenticate each other.

E.g. let us look at what happens when a *conteneur* is taken away:

- The *centre serveur* and the *station* have to authenticate each other.
- The *centre serveur* and the *conteneur* have to authenticate each other.
- The *responsable d'agence* has to be authenticated by the *centre serveur* and by the *conteneur*.
- The *convoyeur* has to be authenticated by the *station* and by the *conteneur*.

Schematically:

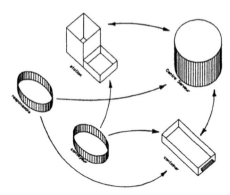

Something similar happens at the moment of arrival as can be seen on the next page.

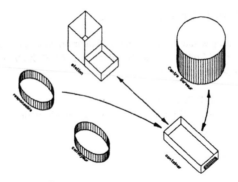

All the special events in the transport cycle of a *conteneur* may require security transactions between 2 or more parties of the list above.

Some events involve an interaction with the *centre serveur*, e.g.:

- *authentification de la station d'arrivée*;
- *autorisation d'ouverture*;
- *autorisation de réouverture*;
- *autorisation d'envoi*;
- *autorisation d'enlèvement*.

Other events do not need an interaction with the *centre serveur* but the problem remains: **how can we make 2 parties trust each other while they basically distrust each other ?**

2.1.2 Security solution

To solve this basic authentication problem we use the following mechanisms:

- People (*responsable d'agence, convoyeur*) are authenticated using (dynamic) passwords.
- Electronic devices (*conteneur, centre serveur, station*) are authenticated using Message Authentication Codes.

This guarantees:

- the integrity of the messages exchanged between the *conteneurs/stations* and the *centre serveur*;
- the authenticity of the origin of these messages;
- the authenticity of the people involved.

Note that this does not mean that the confidentiality of the messages is not important at all: because of psychological reasons towards the customers (Banks, PTT, ...), this confidentiality will be garantueed anyway. As a matter of fact, this confidentiality follows directly from the use of the DES-algorithm.

Integrity of messages and authenticity of their origin

The integrity of the messages and the authenticity of their origin is garantueed by the addition of a Message Authentication Code: a MAC based on the ANSI X9.9 Standard. This MAC is calculated using the following elements:

- the Triple Version of the DES-algorithm;
- some secret key (because a Triple DES is used, the length of the keys is 2 x 56 = 112 bit instead of the usual 56 bit).

Authentication of people

People are authenticated through a Dynamic Password (or Challenge/Response) mechanism using Authenticators or Tokens which are PIN protected. The biggest advantage of this mechanism is that the password is different each time the user has to be authenticated; so, monitoring the connection between the user and the device performing the authentication will not reveal any helpful information.

Another big advantage is that one needs to physically possess 'something' (the Token) and also needs to know 'something' (the PIN) in order to be authenticated properly.

2.1.3 Example

As an example the figure below shows which authentications are executed explicitly (the straight lines) and which authentications follow implicitly (the dotted lines):

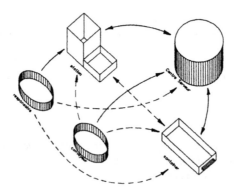

2.2 Logical protection of the centre serveur

The Host Computer itself (actually a STRATUS computer) is located in a physically isolated computer room within the building of the *centre serveur*; basically, this Host Computer takes care of the following functions:

- management of container transports;
- administrative management.

2.2.1 Security needs

The main (logical) component of this Host Computer is a DataBase (actually implemented under SQL2000 or SYBASE).

Outside that computer room (but inside the building of the *centre serveur*), one may find the Key Generator; this is a device connected to the Host Computer and performing all key management functions in the system.

One may find various terminals connected to the Host Computer; these terminals are used for operator functions and may be located all over the building of the *centre serveur*, i.e. inside and outside the computer room.

The configuration described above may be vulnerable to some threats:

- violation of the Key Generator;
- violation of the link between the Host Computer and the Key Generator;
- violation of the logical security of the Host Computer and its DataBase.

2.2.2 Security solution

The Key Generator

The logical (and physical) protection of the Key Generator (built around a Personal Computer) consists of:

- boot protection;
- operator authentication;
- segregation of basic Key Generator functions and management functions;
- disk encryption.

The link between the Key Generator and the Host Computer

The link between the Host Computer and the Key Generator is protected using Line Encryptors.

The Host Computer and its DataBase

When a user logs on to the Host Computer, his identity is verified through the same Dynamic Password mechanism that is used to authenticate a *responsable d'agence*. Note that from now on we will use the expression "operator" instead of "user".

Once the operator has been authenticated through the mechanism described above, he is controlled by VOS (Virtual Operating System, i.e. the operating system of the STRATUS computer).

VOS may identify the operator as having one of the following profiles:

- Super User: this is the highest authority on a STRATUS computer;
- Non-Privileged User: these are ordinary users, having only access to a limited set of VOS-functions;
- Privileged User: these are ordinary users, having access to a more complete set of VOS-functions.

The access of operators having a User profile (Privileged or Non-Privileged) may be restricted to specific:

- directories;
- files (data or programs).

In practice, there will be one person having the status of Super User; this person will be the highest authority in the whole system - therefore, he will also perform the responsibility of Security Officer.

The other operators may have restricted access to some programs depending on their responsibility in the system. E.g., it is possible to exclude an operator from access to SQL2000 (or SYBASE): as a consequence this operator will have no access to the DataBase. Once an operator has gained (legal) access to SQL2000 (or SYBASE), further access is controlled by SQL2000.

In general, SQL2000 controls the DataBase access by:

- operators;
- application programs.

In the practical case of the *centre serveur*, 3 groups for DataBase access have been defined:

- general application programs (e.g. transport management);
- the Crypto-Administrator: this is the software that controls the connection with the Key Generator;
- the operators.

Note that by operators, we mean those operators, to whom DataBase access has been granted by VOS (see above), so we will further use the expression DataBase operators.

A DataBase operator may be one of the following types:

- System Administrator: this is the highest authority for SQL2000;
- DataBase Owner: this is the person being responsible for a specific database (in the case of the Centre Serveur, there is only one DataBase, so the System Administrator and the DataBase Owner have been defined to be the same person);
- an ordinary DataBase operator.

The 2 DataBase object types that need protection, are:

- Tables;
- Stored Procedures.

The access of 'something' (an operator or a program) to a Table is controlled through bit-maps defining the type of access that is allowed:

- Read allowed;
- Update allowed;
- Delete allowed;
- Insert allowed.

The access of 'something' (an operator or a program) to a Stored Procedure is controlled through similar bit-maps defining the type of access that is allowed:

- Execution allowed.

Amongst a lot of other data, the DataBase contains the cryptographic keys of *stations* and *conteneurs*. To protect these cryptographic keys while they are stored in the DataBase, they will be encrypted using the DES-algorithm with some Computer Key. This Computer Key will be program-coded, i.e. it will not be stored on some computer disk. Each time a cryptographic key is needed from the DataBase, it will be decrypted using that Computer Key.

2.3 Key management

2.3.1 General principles

It is very important to emphasize that the cryptographic keys are never transmitted through the communication lines.

At the side of the *conteneur* and the *station*, the cryptographic keys (or the data to generate the cryptographic keys dynamically) are loaded in a secure way into these devices; the cryptographic keys are never transmitted over the microprocessor bus of the *conteneur* or the *station*.

At the *centre serveur* the cryptographic keys are kept in the DataBase. There the protection is twofold:

- the access to the Host Computer is controlled through a Dynamic Password mechanism;
- the cryptographic keys are encrypted using some Computer Key which is program-coded.

2.3.2 Practical implementation

In the AXYTRANS system we need the following cryptographic elements for the *conteneur* and for the *station*:

- MAC Key : A double DES-key used to calculate the MAC (Message Authentication Code) to authenticate a given message; the MAC is calculated using Triple DES, so we need a double DES-key. This key needs to be kept secret; all *conteneurs* and *stations* will have different MAC Keys.

- ENC Key : A double DES-key used to encrypt a given message to protect its confidentiality; the encryption is done using Triple DES, so we need a double DES-key.

This key needs to be kept secret; all *conteneurs* and *stations* will have different ENC Keys.

- Some keys that are used to have a DES based one-way function in the *conteneur* and in the *station*; the keys that are used are single DES-keys that may be public.

- In the case of the *station*, a mechanism to verify the integrity of the MAC Key is needed (and to reload it in a secure way from the *centre serveur* if necessary).

The general hardware architecture (from a cryptographic point of view) of the *conteneur* and of the *station* is depicted below.

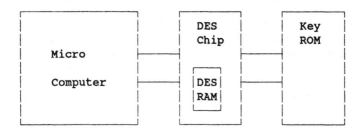

Micro Computer : The microprocessor plus all its program and data storage and interfacing facilities.

DES Chip : The chip used for the fast and secure execution of the DES-algorithm.

DES RAM : Internal volatile storage with a capacity of 4 (single) DES-keys; this storage is only accessible by the DES Chip.

Key ROM : External non-volatile storage with a capacity of 16 (single) DES-keys; this storage is only accessible by the DES Chip.

The *conteneur*

The *conteneur* is a tamperproof device, i.e. it has been physically made impossible to monitor data that are transferred internally between the various components of the *conteneur*. This has to do with the nature of the *conteneur*: if somebody handles the *conteneur* in an unauthorized way, the *conteneur* will physically destroy itself.

So, all cryptographic keys may be stored in the Key ROM without further logical or physical protection; each time a key is needed, the application program of the *conteneur* will 'tell' to the DES Chip which key it has to load from the Key ROM.

In the case of a reset (as the result of a power-down), no special recovery is needed because the Key ROM will preserve its contents.

The *station*

Unlike the *conteneur*, the *station* is not tamperproof. As a consequence, the transfer of the cryptographic keys to the DES Chip may be delicate. Because in the AXYTRANS system, Authenticity is much more important than Confidentiality, especially the MAC Key may be very sensitive.

Therefore, the MAC Key will not be (statically) stored in the Key ROM, but the MAC Key will be (dynamically) generated at the time of installation and immediately stored in the RAM of the DES Chip. Each time the MAC Key is needed, it will not have to be loaded from the Key ROM because it is already available in the DES Chip.

However, because the MAC Key is stored in the volatile DES RAM, a mechanism to detect the probable loss of the MAC Key in case of a reset (as the result of a power-down) is needed. The *station* will then activate a special procedure (the Remote Dynamic Key Loading) to generate the MAC Key, which is entered in the RAM of the DES Chip.

This Remote Dynamic Key Loading enables the *station* to generate its MAC Key dynamically; however this process is based on the principle of Split Knowledge because it is controlled by:

- the *centre serveur*;
- the *responsable d'agence*;
- an *installateur*.

The Key Generator

The Key Generator is a device that is connected to the Host Computer of the *centre serveur*; it is operated by a Key Generator (so, the expression "Key Generator" is used for both the device and the person operating this device).

The main functions of the Key Generator Device are:

- to generate the various cryptographic keys;
- to load the necessary cryptographic keys into the Key ROMs of the *stations* and the *conteneurs*;
- to upload the cryptographic keys to the Host Computer for storage into the DataBase;
- to log all operations.

The access to the Key Generator Device is defined and controlled by the Security Officer; the latter also defines and controls the privileges of the Key Generator Person.

Unix Security & Kerberos

Bart De Decker

K.U.Leuven, Dep. of Computer Science
Celestijnenlaan 200A, B-3001 Heverlee

Abstract. This paper discusses some security issues related to the UNIX operating system, which is today the de facto standard Operating System. The authentication mechanisms have been focused on, both in a central system and in a network environment. It is shown that networking makes UNIX vulnerable if no special measurements are taken. One of these could be the introduction of the Kerberos authentication system which is also becoming a "standard" in open network environments. The Kerberos protocols are described, and their merits and limitations in a possibly hostile environment are discussed.

1   Introduction

Today, the UNIX operating system is the de facto standard operating system. UNIX was not really designed with security in mind [1], which does not mean that UNIX does not provide any security mechanisms at all. However, many of the current UNIX systems have been installed with little or no security enabled. The reasons thereof are mainly historical. UNIX was designed by programmers for programmers and the first sites outside Bell Labs to install UNIX were universities, where there was no real need for security. Today, many businesses and government sites are beginning to install UNIX as well. Hence, UNIX is no longer being used in environments where open collaboration is the primary theme. Moreover, new (networking) features have been added to UNIX over the years, which have expanded its usability but also its vulnerability.

The field covered by this paper is very broad: UNIX, security and Kerberos; hence the aim is not completeness but a better understanding of the basic security mechanisms and an increased awareness of security threats.

The first part takes a close look at the security mechanisms available in UNIX and discusses how they are used in a centralized and in a network environment. In particular, authentication and authorization are explained, and some known security breaches (mostly because of bad management) are brought up or can easily be deduced from the text. We also pay attention to the management problems involved with UNIX.

The second part of this paper focuses on Kerberos, an authentication scheme designed for a –more or less– hostile network environment. Kerberos has been successfully used by several sites and it will be included in a growing number of future UNIX -releases. The protocols that form the basis of Kerberos are unraveled and some of its merits, weaknesses and limitations will be discussed.

2 Unix Security

Unix Security can be subdivided into two main areas of concern. The first deals with *authentication* –both on a centralized system and in a network environment. The purpose of authentication is keeping unauthorized users from gaining access to the system. The second area deals with *protection*. It is concerned with protecting the 'access to the system resources' either by legitimate users or system crackers.

This section is outlined as follows: first, authentication on a centralized UNIX system is discussed; then the protection mechanism is looked at; finally, authentication and protection in a network environment are dealt with.

2.1 Authentication

On a single UNIX system, *user authentication* is based on a password scheme. We assume that the user has an *account* on the system, which involves a login-name, allocated disk-space, and everything that needs to be there for a user to be able to work on the system.

The protocol is very simple:

- first, the user enters his *login-name*,
- secondly, he enters his *password*.

The system encrypts this passwords and compares it with a stored version. If both match, the authentication succeeds and the user is authorized to use the system.

This simple scheme is based on the *secrecy* of passwords. Login-names are usually common knowledge (for instance, they appear in electronic messages) or can be easily guessed, but passwords need to be secrets known only by the user and the system.

The stored version of the password can be found in the password file: /etc/passwd. (A example of this file is shown in Fig. 1.) It contains one line per *identity*. (In the sequel of this paper, we shall use the words *user*, *identity* and *principal* interchangeably.) Each line consists of several fields, separated by semicolons: the login-name (e.g. bob), the encrypted password (e.g. J4wg6qi6IkZEj), the user-identifier (uid) (e.g. 100) and group-identifier (gid) (e.g. 100), some readable information about the user (e.g. & Clark), his home-directory (e.g. /u/bob), and his command interpreter to be invoked after the login (shell) (e.g. /bin/sh). The uid uniquely identifies a user in the system. Note that different login-names may refer to the same uid. In fact, once logged in, there is no way to differentiate between them.

Root, also called the superuser, has a uid equal to zero. This user is – in a sense – very special, since there are no access restrictions for this user (i.e. 'root' is allowed to do everything, even destroying the whole file system).

```
root:Igl6derBr45Tc:0:0:The Superuser:/:/bin/sh
bin:*:1:1:Pseudo-system:/:/bin/sh
bob:J4wg6qi6IkZEj:100:10:& Clark:/u/bob:/bin/sh
alice:ntkYb1ioOkk3j:101:20:& Tyler:/u/alice:/bin/sh
ftp:*:200:20:Anonymous Ftp User:/u/ftp:/bin/sh
...
```

(a) /etc/passwd

```
wheel:*:0:root,operator
bin:*:1:bin
staff:*:10:bob,peter,kate
security:JXk3opPklZ3xR:20:alice,bob,root
...
```

(b) /etc/group

Fig. 1. Files Used in the Authentication Protocol

Principals are assembled into *groups*. The **/etc/group** file contains one line per *group*. Each line consists of several fields also separated by semicolons: the group-name (e.g. **staff**), possibly an encrypted password, the group-identifier (gid) (e.g. **10**) and a list of group-members (e.g. **bob,peter,kate**). Users can be a member of many groups, the password file lists only the primary group.

Many commands use the information stored in **/etc/passwd** and **/etc/group**. Hence, they need to be readable by any process. Note that these files contain encrypted passwords! A modified DES algorithm is used to encrypt passwords: a constant is encrypted with the user's password as key. The algorithm is iterated 25 times to slowdown the encryption process. Moreover a 2-character *salt* (4096 different values) is added to modify the key, the first two characters of the password-field being the salt.

Although passwords cannot be deduced from the stored encrypted entities, and although the presence of a salt makes dictionary attacks difficult to conduct, two observations weaken the password security:

- Processor speeds are increasing dramatically, and many sites have thousands of interconnected machines available that can be used to break passwords.
- UNIX places very little restrictions on what may be used as a password. In [4] the author describes experiments conducted to break user's passwords: almost 24% of the passwords could be cracked by using a limited dictionary and easy-to-guess choices.

On a centralized system, one needs already an account on the system in order to be able to access the password file, but one can always try to crack the password of a more privileged user, such as 'root'. In a network environment, however, this need not to be true: many sites allow 'anonymous file transfer'; special precautions should be taken to prevent anonymous users from accessing /etc/passwd (see 2.3.1). Also, breaking a password on a local system may open doors to other systems.

Therefore, some UNIX versions have removed password information from the password file (/etc/passwd) and have stored it in a shadow file which is only accessible to 'root'.

To make the system less vulnerable, the standard *passwd* command, which changes the user's password, should be replaced by a newer version that prevents the user from choosing an easy-to-guess password. Guidelines for selecting passwords can be found in [3].

Figure 2 shows the two processes and three programs involved in the login procedure. (Straight boxes represent processes, dashed boxes represent the programs these processes execute.)

The *init* process starts a *getty* process for every terminal port that is turned on. *Getty* reads the login-name, executes the *login* program with the login-name as argument. *Login* requests the password, validates the (login-name, password) pair against those stored in the /etc/passwd file, and executes the user's command interpreter *sh*. The *sh* reads user's commands and executes them.

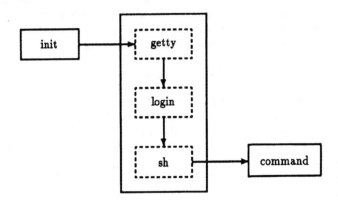

Fig. 2. Processes Involved in the Login Procedure

Apart from preventing unauthorized users from gaining access to the system, the login procedure assigns the user's uid and gid(s) to the command interpreter (*sh*). These ids will be inherited by all the processes that are directly or indirectly created this interpreter. Uid and gid(s) are used to determine the access rights to system resources (such as files).

2.2 Authorization

Every UNIX *file* has a true owner and one group owner. The true owner can modify the *attributes* of the file. One of this attributes is a limited *access control list* (acl) which determines what operations are allowed by (a) the owner, (b) group-members and (c) the others. UNIX distinguishes between read, write and execute access; hence, nine bits suffice for the acl.

Ownership of *processes* is more complex. UNIX associates four numbers with each process: two *real* numbers (ruid and rgid), and two *effective* numbers (euid and egid). The real numbers are used for accounting purposes, while the effective numbers determine the access-privileges of the process to system resources. Usually, real and effective numbers are the same and equal to the uid and gid of the user that created the process. When a process creates (spawns) a new child process, the latter inherits the real and effective numbers of its parent.

Two additional attributes are provided for each file: the *set user-id* (suid) and the *set group-id* (gid) attributes. When set, the process that executes the program stored in the file changes its privileges: the euid (resp. egid) becomes the owner (resp. group owner) of the file.

A process can also switch the effective and real numbers. Moreover, a process with euid equal to zero (i.e. 'root-privileges') can change real and effective numbers to any number. For example, *getty* and *login* execute with root-privileges; before executing the *sh*, *login* changes the uids and gids to those of the user logged in. Hence, except for programs with suid or sgid attributes, all processes created by the *sh* will execute with the user's privileges.

The authorization protocol works as follows:

- if the euid represents 'root', access is allowed;
- otherwise, if the euid equals the owner of the file, then access is allowed if the operation is permitted to the file-owner;
- otherwise, if the egid equals the group owner of the file, then access is allowed if the operation is permitted to the group owner;
- otherwise, access is allowed if the operation is permitted by the others-field of the acl.

Changing the privileges of a process is common in UNIX. An example is the password program, which changes the user's password. The process executing this program has to alter the /etc/passwd file. However, the system cannot allow this file to be world-writable, since any user could then change passwords at will with an editor; hence the acl allows only the owner (root) to write this file, and everyone (owner, group owner and others) to read it. When the user gives the 'passwd' command to the *sh*, the latter will spawn a new process (with real and effective ids inherited from *sh*). When this new process executes the

/bin/passwd program, the euid is changed to zero, since /bin/passwd is owned by root and has the suid attribute set. The process has now root privileges and is allowed to update /etc/passwd.

2.3 Authentication in a Network Environment

UNIX systems can be linked together to form networks. Over the years, many networking features have been added to UNIX systems, which have –without any doubt– increased the usability of UNIX, but also made it far more difficult to control.

In this section, a distinction has been made between autonomous systems, collaborating systems and integrated systems, the distinctive characteristic being *trust*: "How much does the local site trust the other (remote) sites?"

Autonomous Sites. Autonomous Systems do not rely on *trust*. Typically, the interconnected systems exchange information after an authentication procedure that resembles very well the login procedure: i.e. a login-name and a password have to be presented before access to the remote system is allowed.

Systems connected via direct or dial-up links provide the *uucp*-class of programs, which allow for file transfer and remote execution. Requests are usually spooled, and a set of specific security mechanisms is available such as callback, limited access to only a subtree of the file system, etc. These restrictions are specified in a set of configuration files.

On the other hand, *Internet* access is available via *telnet* (a terminal emulator protocol) and via *ftp* (a file transfer protocol). These programs are not unsafer than 'login', however, passwords are sent in cleartext through the network and might be captured by possible intruders.

Ftp also provides an 'anonymous' login possibility, which allows users with no account on the system to login as an anonymous user. It is the primary way software and documents are distributed on the Internet. The ftp-daemon, which makes it all work, will make files outside the home-directory of 'ftp' (/u/ftp in Fig. 1.a) invisible and inaccessible. Note that the /etc should not be a subtree of that directory; otherwise, anyone can copy /etc/passwd and try to break the encrypted passwords.

For diskless workstations, a trivial file transfer protocol *tftp* has been designed which does not require authentication. Here too, the daemon process that services tftp-requests should not allow access to the complete filesystem (i.e. it should be started with the 'secure' option set on, and /etc should not belong to the accessible subtree).

Collaborating Systems. The second class of networked systems are called *collaborating systems*. They are usually linked through a local area network (LAN). (See Fig. 3 as an example.)

Collaborating systems provide the means to share system resources such as printers or files. Usually, users have accounts on many of these systems and do

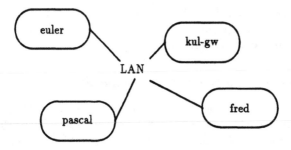

Fig. 3. A Local Area Network

not want to authenticate themselves with passwords each time they access a resource on a remote system.

The programs involved in this kind of collaboration are the so-called *remote*-variants of normal UNIX-commands such as rcp, rsh, rlogin, etc., available in the Berkeley (and derived) UNIX networking software.

The software allows the specification of other hosts (and possibly users) who are considered *trusted*. This means that remote logins and remote command execution will be permitted without requiring the user to enter a password. This is convenient, however, also very insecure. The internet worm [5] made excessive use of this mechanism to invade other 'trusted' systems.

Figure 4 shows some samples of /etc/hosts.equiv and $HOME/.rhosts files. The first one lists the trusted hosts. If a users tries to log in (using rlogin) or execute a command (using rsh) remotely from one of the systems in /etc/hosts.equiv, and that user has an account on the local system with the same login-name, access is permitted without requiring a password. The $HOME/.rhosts variant is similar but allows trusted access only to specific host-user combinations instead of hosts in general.

Integrated Systems. Integrated Systems base their communication on *remote procedure calls*. An example is the Network File System protocol[10] (NFS), which is designed to allow several hosts to share files over the network. It allows for diskless workstations to be installed in the offices, while keeping all disk storage in a central location.

/etc/exports lists the file systems that are exported by NFS and can thus be mounted by trusted remote systems. Figure 5 shows an excerpt from this file. The 'root=' keyword specifies the hosts that are allowed to have 'root-access' to the files in the named file system. The 'access=' keyword lists the hosts that

```
euler
pascal
...
```

(a) /etc/hosts.equiv

```
kul-gw
pascal alice
```

(b) /u/bob/.rhost

Fig. 4. Files Expressing Trust (of 'fred')

are allowed to mount the file system. Note, however, that usually, the mount protocol does not verify the identity of the mounting hosts.

```
/usr
/export/root/fred      -access=fred
/spec/alice            -root=alice
```

Fig. 5. /etc/exports

In the remote procedure call, the caller authenticates to the callee, but the fields (such as uid, gid, hostid, etc.) are put in cleartext in the request message. Hence, it is not very difficult to inject requests with a false identity.

Sun proposed a *secure* remote procedure call protocol. It uses DES encryption and public key cryptography to authenticate both users and hosts in the network. DES is used to encrypt timestamps, public key encryption (PKE) provides a safe way to distribute a secret DES-key.

The PKE scheme uses the Diffie-Hellman method. Every principal has a public key (PK) and a secret key (SK), where $PK = \alpha^{SK}$ modulo some large constant. If two principals 'x' and 'y' need to share a key, they use key K_{xy} where $K_{xy} = PK_x^S K_y = PK_y^S K_x$. Since the secret keys are only known to the principals themselves, nobody else but principals 'x' and 'y' can calculate K_{xy}. In the sequel of this paper, the notation $\{info\}_K$ means 'info' encrypted with key K.

Figure 6 sketches the protocol. When a client wishes to talk to a server, he generates a random key to be used for encrypting timestamps. This key is called

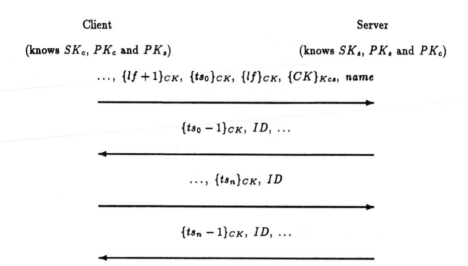

Fig. 6. Secure Remote Procedure Call

the *conversation key*, CK. The client encrypts CK using a public key scheme described above and sends it to the server in the first request message. This is the only public key encryption in this protocol. All further encryptions use DES with CK as key. In the first request-messages, the client sends also: its identity (*name*), the lifetime of CK (lf), a verifier for this lifetime-field (i.e. $lf - 1$) and a verifier for his identity (i.e. a timestamp ts_0), all three encrypted with CK.

The server process decrypts $\{CK\}_{Kcs}$ to get hold of CK. Then, he checks and verifies the lifetime and the identity of the client.

In the reply message, the server proves his identity by sending the client's timestamp minus one ($ts_0 - 1$), encrypted with CK. Since only client and server know CK, only the server would be capable to generate this cryptographic token. An identifier (ID) is also returned for future use by the client.

The last two messages in this protocol show the authentication info exchanged in further request-reply messages: the client sends its ID and proves it through a verifier (an encrypted timestamp ts_n). The verifier also counters play-back attacks, since the server always demands timestamps that are greater than the ones seen before. Again, the server completes the mutual authentication by returning the client's timestamp minus one ($ts_n - 1$).

Public and (encrypted) secret keys are stored in Yellow Pages. At login time, the user's secret key is decrypted and stored in a secure local keyserver, who does all the public key encryption. It is not very fast, and takes roughly 1 second on a Sun-3 machine to perform this operation. However, public key encryption is only necessary in the first two messages of a transaction. Moreover, the keyserver caches results of previous computations so that it does not have to recompute the exponential every time.

2.4 UNIX Administration

The most difficult part of UNIX is 'administration'. As should be clear from the examples given above, many files determine the security behaviour of the system. We have not dealt with all of them. Too many tools have their own configuration files, in which one can express access restrictions, or select extra security measures. For example, the *uucp* tools, *ftp*, *nfs*, etc. have their own rules for preventing unauthorized access. It takes a real expert, and a good amount of experience to manage it all –in a responsible way. There is not yet a 'security tool' that coordinates the different actions that have to be taken. Moreover, there is really a need for standardization and simplification, so that new tools would not introduce yet another set of files to manage.

Fortunately, public domain packages, such as 'Cops'[12], that help the administrator check for security holes, become available. They are not 'the' solution to the administration problem UNIX clearly suffers, but may help in anticipation of a real management tool.

3 The Kerberos Authentication System

The Kerberos Authentication System[6, 8] was initially introduced by MIT to meet the needs of Project Athena. It has since been adopted by a number of other organisations for their own purposes, and is being discussed as a possible standard.

This section describes the environment for which Kerberos has been designed, describes the protocols used and looks at its strengths and weaknesses in a UNIX context.

3.1 Introduction

Project Athena's computing environment consists of a large number of workstations –many of them installed in publicly accessible places– and a smaller number of large autonomous server machines. The servers provide file storage, mail services, print spooling and perhaps some computing power; the workstations are mostly used for interaction and computing. The workstations may have local disks, but these are not used for holding user data. Moreover, they are not physically secure and anyone could remove or alter any portion of the disk. Kerberos is used to authenticate users that work on these workstations. They need access to their personal files stored on the server machines, or to remote printers.

A more complete picture of the Athena environment can be found in [11].

3.2 The Kerberos Protocols

The Kerberos authentication system consists of three different protocols, which are merged together. In this subsection we will disentangle them and discuss them in turn:

1. the *authentication protocol*, in which two parties sharing a secret (key) prove their identity (one-way or mutually);
2. the *key-distribution protocol*, which is used to transfer safely a secret (key) between two distant parties;
3. the *single sign-on protocol*, which converts a weak secret (password) into a strong secret (random bits).

Another approach for explaining the rationale behind the protocols can be found in [7], where the security threats and solutions are presented in a dialogue.

The Authentication Protocol. The authentication protocol is based on *time* and encryption using a *secret key*.

The scenario is depicted in Fig. 7. Suppose the client 'Alice' wants mutual authentication with the server 'Bob'. Both share (i.e. know) the secret key K_{ab}.

Alice sends an *authenticator* (A_a) to Bob, that contains her name, possibly her address, and a timestamp (the time on her local clock). The authenticator is sealed with the secret key K_{ab}.

When Bob receives the authenticator, he decrypts it, checks Alice's name and the source-address of the message (this should be the same as the address included in the authenticator). Finally, Bob verifies the 'freshness' of the message by comparing the included timestamp with his local clock (and possibly with the timestamp included in previous authenticators from Alice). If all tests succeed, Bob can be sure that the authenticator originated from Alice, and that it is not a replay.

If mutual authentication is wanted, Bob will return a function of Alice's timestamp encrypted with the secret key.

The Key Distribution Protocol. The second problem to solve is *how to make two principals share a secret key*. Suppose the client Alice wants to authenticate with server Bob, but they do not share a secret key. Hence, the previous protocol cannot be applied. However, if Alice had a means to transfer safely the secret key together with the authenticator, the problem would be solved. Therefore, Alice will ask a *trusted third party* (The Key Distribution Center or KDC) to create a secret key for her to share with Bob. The KDC shares a secret key with all the principals. This key is called the principal's *master key*. This scheme is much more economical since –in the case of n principals– only n master keys are needed, as opposed to the $O(n^2)$ keys if all principal-pairs need to share a key. The KDC will create a temporary key, also called a *session key*, on request, and return two sealed packets, both containing this key. The first packet, also called *certificate*, is meant for the requesting client and will be encrypted with his master key; the second packet, called a *ticket*, is destined for the server, and hence, will be encrypted with the server's master key.

To summarize the key distribution protocol: a client requests the KDC a certificate and a ticket, both containing a new secret session key. From the certificate, the client can extract this session key. The ticket will be sent to the server to notify him of this session key.

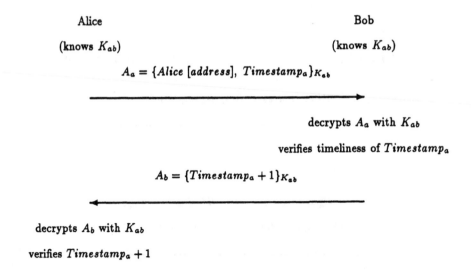

Fig. 7. The Authentication Protocol

The scenario is depicted in Fig. 8. Client Alice sends a request (in cleartext) to the KDC, containing her name *(Alice)*, possibly her address(es), the server's name *(Bob)*, a nonce (N_a) which is a random number and the lifetime for the session key *(lifetime)*.

The KDC will return to Alice a certificate (C_a^b) and a ticket (T_b^a). The certificate is encrypted with Alice's master key (K_a) and contains Alice's nounce, the session key (K_{ab}), the name of the server *(Bob)* and the lifetime for the session key *(lifetime')*, which may be less than the requested lifetime. Since Alice knows her master key, she can decrypt the certificate, check nounce, server's name and lifetime and store the session key together with the ticket for future use. (The nounce was added to the protocol to counter replay attacks; the address(es) add some extra security, since the server can reject stolen tickets when they are sent from the wrong address.)

The ticket is meant for the server Bob, and is encrypted with Bob's master key he shares with the KDC. It contains the session key (K_{ab}), the client's name *(Alice)*, possibly the client's address(es), and the lifetime for the session key *(lifetime')*.

If this protocol is combined with the authentication protocol, then the client will send not only an authenticator to the server but also a ticket, from which the server can extract the session key used to encrypt the authenticator. However, the server needs to verify the validity of the ticket (it should contain the client's name, come from the right address and have not been expired yet).

Without knowledge of the session key, stolen tickets are useless since the intruder will not be able to construct the right authenticator. Moreover, even if the session key has been exposed to an intruder, the latter cannot misuse the ticket very long due to included lifetime.

Alice   KDC

(knows K_a)   (knows K_a and K_b)

$$Alice\ [address],\ Bob,\ N_a,\ lifetime,\ \ldots$$

$$\longrightarrow$$

Creates K_{ab}

$$C_a^b = \{N_a,\ K_{ab},\ Bob,\ lifetime',\ \ldots\}_{K_a},$$

$$T_b^a = \{K_{ab},\ Alice\ [address],\ lifetime',\ \ldots\}_{K_b}$$

$$\longleftarrow$$

decrypts C_a^b with K_a

Fig. 8. The Key Distribution Protocol

Since the KDC issues tickets, it is called the *Ticket Granting Server* or TGS. In the sequel of this paper, we shall use KDC and TGS interchangeably.

The Single Sign-On Protocol. In the previous subsection, we stated that every principal in the system needs to share a key with the TGS. This is fine for non-human principals, such as server processes, but impractical for human users. They want to share an easy-to-remember key (such as a password), but keys based on passwords are known to be weak.

The *single sign-on protocol* is meant to convert such weak keys into strong once. Again, a *trusted third party* is needed, which is called the *authentication server* or AS. All human principals share a secret password with the AS. The protocol is very similar to the key distribution protocol. Only now, the client wants to share a strong key with the TGS. Hence, it is sufficient to substitute 'server' by 'TGS' and to send the request to the AS instead of the TGS in the previous protocol. The AS will return a certificate and a ticket. The first one is encrypted with a key derived from the password (K_{pw-a} in Alice's case) and contains amongst others a strong key to share with the TGS. The ticket is encrypted with the TGS's master key, and is called the *ticket-granting-ticket* or *tgt*, since a client can request tickets for servers with this ticket. The scenario is depicted in Fig. 9.

If this protocol is combined with the key distribution protocol, then a client does not share a master key with the TGS. Instead, the request message sent to

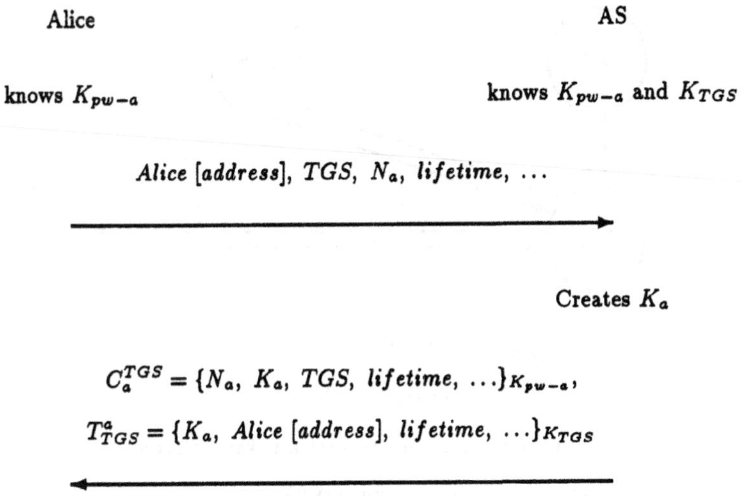

Alice   AS

knows K_{pw-a} knows K_{pw-a} and K_{TGS}

Alice [address], TGS, N_a, lifetime, ...

————————————————————————————————▶

Creates K_a

$$C_a^{TGS} = \{N_a,\ K_a,\ TGS,\ lifetime,\ ...\}_{K_{pw-a}},$$

$$T_{TGS}^a = \{K_a,\ Alice\ [address],\ lifetime,\ ...\}_{K_{TGS}}$$

◀————————————————————————————————

Fig. 9. The Single Sign-On Protocol

the TGS will contain the *tgt*, previously obtained from the AS, and an authenticator that proves that the ticket has not been stolen.

The Complete Scenario. In Kerberos, all three protocols have been merged together:

1. at login time, the client requests a *tgt* from the AS,
2. with that *tgt*, the client can request a ticket for a server he wants to authenticate with,
3. this ticket is sent to the server to provide him with a session key, together with an authenticator to prove the client's identity.

The scenario is depicted in Fig. 10.

3.3 Multiple Domains

Kerberos does not require that all principals are registered with the same central authority (i.e. AS/TGS). Instead, multiple domains, called *realms*, are provided. Each realm has its own AS and TGS. Principals are registered in one realm and can authenticate themselves with principals registered in other realms, as long as a trusted chain of ticket-granting servers that link the two realms can be found.

The Kerberos protocol is easily extended to a multiple domain environment. The client, registered in realm X, cannot obtain from his TGS (say TGS_X) a ticket for a server registered in realm Y, since the server and TGS_X do not share

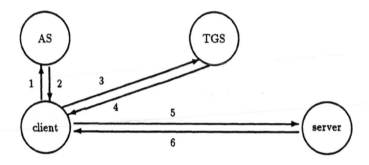

1: request a tgt for TGS

2: certificate, tgt (for TGS)

3: tgt, authenticator, request a ticket for server

4: certificate, ticket (for server)

5: ticket, authenticator, ...

6: authentication-reply, ...

Fig. 10. The Complete Scenario

a key. Instead, the client asks his TGS_X for a *tgt* for the next TGS in the trusted chain, whom he asks a *tgt* for the following TGS and so on. Finally, the client will have a *tgt* for the TGS of realm Y (TGS_Y), whom he can ask a ticket for the server. With that ticket, the client can go directly to the server.

A simplified scenario (with only two realms involved) is shown in Fig. 11. The first two exchanges (1 and 2) have been omitted since they occur only at login time.

3.4 Kerberos: Merits, Weaknesses and Limitations.

Installing Kerberos on a network of UNIX systems will certainly enhance the security of the network since passwords are never sent in cleartext.

From the user's point of view, little will change. In fact, he will hardly notice the presence of Kerberos. The ticket-granting ticket is obtained from the authentication server as part of the login process, and the tickets are automatically destroyed when the user logs out.

However, if the user's login session lasts longer than the lifetime of the ticket-granting ticket (approximately 8 hours), applications that use Kerberos authentication will fail, and the user will have to fetch a new ticket-granting ticket thereby re-entering his password.

Applications using the Kerberos authentication protocol will have to be adapted. Extra calls to the Kerberos library routines have to be inserted in

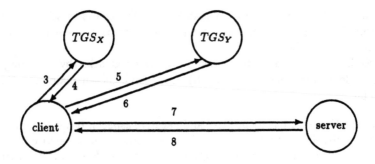

3: tgt-x, request a tgt for TGS_Y

4: certificate, tgt-y (for TGS_Y)

5: tgt-y, authenticator, request a ticket for server

6: certificate, ticket (for server)

7: ticket, authenticator, ...

8: authentication-reply, ...

Fig. 11. Kerberos and Multiple Domains

the source code. Note that Kerberos will only increase the security for these applications.

Kerberos works fine in the environment for which it has been designed, i.e. single-user anonymous workstations that need to access remote resources provided by (secure) server machines. Apart from some local temporary storage and computer power, these workstations have no own resources, and certainly none they want to share with others. This is in contrast with a typical UNIX network environment, where every host has its own resources that may be shared with others. Moreover, many users can be concurrently logged in on such a host.

One problem is the caching of session keys and tickets in the file system. On a single-user system with local disk space, this works fine, but on a multi-user system, keys might be exposed. It is even worse with diskless workstations, where storing and retrieving keys would mean that they travel (in cleartext) through the network.

The protocol itself has also a few weaknesses. Typically, authenticators have a lifetime of 5 minutes, which could allow replay attacks if an intruder captured a ticket and matching authenticator. A server can only recognize a replay if all previous 'live' authenticators are kept. But it is doubtful that many servers will or can do.

Kerberos is based on synchronized clocks. This means that there must be an authenticated time service available, otherwise clocks could be set and reset at will. It seems contradictory to base an authentication scheme on another authentication scheme.

Kerberos suffers also from password-guessing attacks (by recording and analyzing login dialogues with the AS), from some chosen plaintext attacks. See also [9].

Clearly, Kerberos has its positive and negative aspects. It has been designed for a very particular environment, not the most common UNIX environment, and is mostly used for user-server authentication. It still suffers some weaknesses that could be overcome by a challenge-response authentication scheme, although the latter would introduce an extra message to be sent, which cannot easily be fit in the 'remote procedure call' semantics used by many applications.

4 Conclusions

UNIX was not designed with security in mind and many sites have installed UNIX with little security measures enabled. However, UNIX provides sufficient means to secure a system, but the administration is too complicated. Many files influence the security-behaviour of the system, and changing one line or even one parameter can make the difference between a secure system and a system open for crackers. Luckily, public domain packages that help to look for open doors, become more and more available and hopefully will grow towards a real security management tool.

In a network environment, authentication schemes where passwords travel in cleartext are to be banned, especially if the environment is open and possibly hostile. In this respect, the Kerberos authentication scheme –although not perfect– certainly enhances the security. Its major drawbacks are the necessity for synchronized clocks, the not negligible encryption overhead and the fact that it is not yet well integrated in UNIX. Although better (nounce-based) solutions might become available in the future, Kerberos is available today, and can be freely obtained from MIT.

References

1. Ritchie, Dennis M., "On the Security of UNIX." May 1975. Reprinted in UNIX *System Manager's Manual*, 4.3 Berkeley Software Distribution. University of California, Berkeley. April 1986.
2. Patrick H. Wood, Stephen G. Kochan, "UNIX System Security", *Howard W. Sams & Company*, 1985.
3. David. A. Curry, *"Improving the Security of Your UNIX System,"* SRI International Tech. Report ITSTD-721-FR-90-21, April 1990.
4. Daniel V. Klein, *"Foiling the Cracker": A Survey of, and Improvements to, Password Security*, Draft, 1990
5. Eugene H. Spafford, "The Internet Worm Program: An Analysis," Purdue Tech. Report CSD-TR-823, November 1989, 1988.

6. J.G.Steiner, B.C. Neuman, and J.I. Schiller, "Kerberos: An Authentication Service for Open Network Systems," In *Proc. Winter USENIX Conference*, Dallas, pp. 191-202, February, 1988.

7. Bill Bryant, *"Designing an Authentication System: a Dialogue in Four Scenes,"* Draft, February 8, 1988.

8. John Kohl, Clifford Neuman, *The Kerberos Network Authentication Service*, MIT project Athena, RFC draft #4, December 20, 1990.

9. Steven M. Bellovin, Michael Merritt, "Limitations of the Kerberos Authentication System," in *Proc. Winter USENIX Conference*, Dallas, 1991.

10. R. Sandberg, D. Goldberg, et al., "Design and Implementation of the Sun Network Filesystem," in *Proc. Summer USENIX Conference*, 1985.

11. G.W. Treese, "Berkeley Unix on 1000 Workstations: Athena Changes to 4.3BSD," in *Proc. Winter USENIX Conference*, 1988.

12. Dan Farmer, "COPS and Robbers, UN*X System Security", January 1991. Available from many *Internet archive sites*.

Author Index

Springer-Verlag
and the Environment

We at Springer-Verlag firmly believe that an international science publisher has a special obligation to the environment, and our corporate policies consistently reflect this conviction.

We also expect our business partners – paper mills, printers, packaging manufacturers, etc. – to commit themselves to using environmentally friendly materials and production processes.

The paper in this book is made from low- or no-chlorine pulp and is acid free, in conformance with international standards for paper permanency.

Lecture Notes in Computer Science

For information about Vols. 1–665
please contact your bookseller or Springer-Verlag

Vol. 703: M. de Berg, Ray Shooting, Depth Orders and Hidden Surface Removal. X, 201 pages. 1993.

Vol. 704: F. N. Paulisch, The Design of an Extendible Graph Editor. XV, 184 pages. 1993.

Vol. 705: H. Grünbacher, R. W. Hartenstein (Eds.), Field-Programmable Gate Arrays. Proceedings, 1992. VIII, 218 pages. 1993.

Vol. 706: H. D. Rombach, V. R. Basili, R. W. Selby (Eds.), Experimental Software Engineering Issues. Proceedings, 1992. XVIII, 261 pages. 1993.

Vol. 707: O. M. Nierstrasz (Ed.), ECOOP '93 – Object-Oriented Programming. Proceedings, 1993. XI, 531 pages. 1993.

Vol. 708: C. Laugier (Ed.), Geometric Reasoning for Perception and Action. Proceedings, 1991. VIII, 281 pages. 1993.

Vol. 709: F. Dehne, J.-R. Sack, N. Santoro, S. Whitesides (Eds.), Algorithms and Data Structures. Proceedings, 1993. XII, 634 pages. 1993.

Vol. 710: Z. Ésik (Ed.), Fundamentals of Computation Theory. Proceedings, 1993. IX, 471 pages. 1993.

Vol. 711: A. M. Borzyszkowski, S. Sokołowski (Eds.), Mathematical Foundations of Computer Science 1993. Proceedings, 1993. XIII, 782 pages. 1993.

Vol. 712: P. V. Rangan (Ed.), Network and Operating System Support for Digital Audio and Video. Proceedings, 1992. X, 416 pages. 1993.

Vol. 713: G. Gottlob, A. Leitsch, D. Mundici (Eds.), Computational Logic and Proof Theory. Proceedings, 1993. XI, 348 pages. 1993.

Vol. 714: M. Bruynooghe, J. Penjam (Eds.), Programming Language Implementation and Logic Programming. Proceedings, 1993. XI, 421 pages. 1993.

Vol. 715: E. Best (Ed.), CONCUR '93. Proceedings, 1993. IX, 541 pages. 1993.

Vol. 716: A. U. Frank, I. Campari (Eds.), Spatial Information Theory. Proceedings, 1993. XI, 478 pages. 1993.

Vol. 717: I. Sommerville, M. Paul (Eds.), Software Engineering – ESEC '93. Proceedings, 1993. XII, 516 pages. 1993.

Vol. 718: J. Seberry, Y. Zheng (Eds.), Advances in Cryptology – AUSCRYPT '92. Proceedings, 1992. XIII, 543 pages. 1993.

Vol. 719: D. Chetverikov, W.G. Kropatsch (Eds.), Computer Analysis of Images and Patterns. Proceedings, 1993. XVI, 857 pages. 1993.

Vol. 720: V.Mařík, J. Lažanský, R.R. Wagner (Eds.), Database and Expert Systems Applications. Proceedings, 1993. XV, 768 pages. 1993.

Vol. 721: J. Fitch (Ed.), Design and Implementation of Symbolic Computation Systems. Proceedings, 1992. VIII, 215 pages. 1993.

Vol. 722: A. Miola (Ed.), Design and Implementation of Symbolic Computation Systems. Proceedings, 1993. XII, 384 pages. 1993.

Vol. 723: N. Aussenac, G. Boy, B. Gaines, M. Linster, J.-G. Ganascia, Y. Kodratoff (Eds.), Knowledge Acquisition for Knowledge-Based Systems. Proceedings, 1993. XIII, 446 pages. 1993. (Subseries LNAI).

Vol. 724: P. Cousot, M. Falaschi, G. Filè, A. Rauzy (Eds.), Static Analysis. Proceedings, 1993. IX, 283 pages. 1993.

Vol. 725: A. Schiper (Ed.), Distributed Algorithms. Proceedings, 1993. VIII, 325 pages. 1993.

Vol. 726: T. Lengauer (Ed.), Algorithms – ESA '93. Proceedings, 1993. IX, 419 pages. 1993

Vol. 727: M. Filgueiras, L. Damas (Eds.), Progress in Artificial Intelligence. Proceedings, 1993. X, 362 pages. 1993. (Subseries LNAI).

Vol. 728: P. Torasso (Ed.), Advances in Artificial Intelligence. Proceedings, 1993. XI, 336 pages. 1993. (Subseries LNAI).

Vol. 729: L. Donatiello, R. Nelson (Eds.), Performance Evaluation of Computer and Communication Systems. Proceedings, 1993. VIII, 675 pages. 1993.

Vol. 730: D. B. Lomet (Ed.), Foundations of Data Organization and Algorithms. Proceedings, 1993. XII, 412 pages. 1993.

Vol. 731: A. Schill (Ed.), DCE – The OSF Distributed Computing Environment. Proceedings, 1993. VIII, 285 pages. 1993.

Vol. 732: A. Bode, M. Dal Cin (Eds.), Parallel Computer Architectures. IX, 311 pages. 1993.

Vol. 733: Th. Grechenig, M. Tscheligi (Eds.), Human Computer Interaction. Proceedings, 1993. XIV, 450 pages. 1993.

Vol. 734: J. Volkert (Ed.), Parallel Computation. Proceedings, 1993. VIII, 248 pages. 1993.

Vol. 735: D. Bjørner, M. Broy, I. V. Pottosin (Eds.), Formal Methods in Programming and Their Applications. Proceedings, 1993. IX, 434 pages. 1993.

Vol. 736: R. L. Grossman, A. Nerode, A. P. Ravn, H. Rischel (Eds.), Hybrid Systems. VIII, 474 pages. 1993.

Vol. 737: J. Calmet, J. A. Campbell (Eds.), Artificial Intelligence and Symbolic Mathematical Computing. Proceedings, 1992. VIII, 305 pages. 1993.

Vol. 738: M. Weber, M. Simons, Ch. Lafontaine, The Generic Development Language Deva. XI, 246 pages. 1993.

Vol. 739: H. Imai, R. L. Rivest, T. Matsumoto (Eds.), Advances in Cryptology – ASIACRYPT '91. X, 499 pages. 1993.

Vol. 740: E. F. Brickell (Ed.), Advances in Cryptology – CRYPTO '92. Proceedings, 1992. X, 565 pages. 1993.

Vol. 741: B. Preneel, R. Govaerts, J. Vandewalle (Eds.), Computer Security and Industrial Cryptography. Proceedings, 1991. VIII, 275 pages. 1993.

Vol. 742: S. Nishio, A. Yonezawa (Eds.), Object Technologies for Advanced Software. Proceedings, 1993. X, 543 pages. 1993.